An Approach to Thinking: A Case Study About Water Quality Assessment

D. P. Miller

Intentionally blank

Contents

1 Introduction   4
2 Review Precedents   13
3 Start Thinking   15
4 Begin Understanding   17
5 Probability and Bayes Rule   21
6 Making Choices   27
7 The Details to Get Set Up from Scratch   32
8 Computer Code for an ASCII Demo   42
9 Other Options   46
10 Reasonable Expectations   55
11 Choose the Programming Language   65
12 How Well Does Heap Sort Work?   79
13 Looking Inside Files   87
14 Merge   91
15 Imperfect Data and Bayes Theorem   97
16 Probability Demo   100
17 Translators   118

CONTENTS

18 Two Databases	122
19 The Zipper	126
20 Watersheds and Finding a Segment	148
21 Where is Standard Deviation?	161
22 Do It	167

Chapter 1

Introduction

This book is about thinking, and preliminaries for the reader include: who are the intended audience, what are the themes, what will become useful or hopefully at least helpful.

The intended audience is a narrow slice through modern society, including the precocious middle schooler, the high schooler who might be interested, the college student who wants to know something about computers and using computers but is majoring in a subject that will not provide an in-depth peek, the grad student who now needs some working knowledge of computers, the recently hired professional in a non-technical position who soon learns that some knowledge about computer usage is actually necessary, the MBA who has worked up into a decision making position and realizes that a modest understanding of computer capabilities is needed to make good decisions, the technical professional who wishes to apply computers in ways apparently not available for one reason or another, and, the curious person who is intrigued by computers but never had the opportunity to take the dive. Why would someone want to read this book? To increase their own skills and knowledge about how an individual may use a computer.

However, more important than why is how. Now the intended audience is a wide slice through modern society, those who because of cost or access, are or believe they are unable to pursue computer literacy. For these people, a way is shown here that is about as cheap as possible, and with access about as free and available as possible, and, tailored for the inexperienced.

In addition to computer literacy, the following themes are touched upon: assessment, quality, exclusive privileged access, self-sustaining independence, frugality and simplicity, truth and understanding, and, ex-

CHAPTER 1. INTRODUCTION 5

ploring one limited but detailed example of thinking and computer coding.

This book contains a case study, the idea explored is thinking about water quality assessment. The thought process starts with exploring how water quality assessment happens in the real world, reflects on the idea of doing assessment differently, discusses how this reflection actually involves thinking by a human mind combined with thinking assisted by computer technology, and explores some capabilities of computer technology to design and validate an alternative methodology for assessment. Useful and helpful outcomes will be derived when the readers think about, and extend, the ideas and approaches alluded to here, which are thinking augmented by computer technology. This case study is narrowly based on water quality assessment, which may or may not have any relevance about any other thinking. This case study is about the process, which hopefully does have relevance about acquiring computer literacy and why it is important.

Computer code is a significant part of this book, meant to be read rather than skimmed over. The coding style here is unpolished but attempts to be clear and simple, while at the same time forced to fit on a page of this book. Code is not narrative and code for readers with little or no previous experience is a slow read. How much of the code and how thoroughly the code is read will depend on the reader and what the reader wants to get out of this book. All the code are intended to be complete, that is, work as is and execute as intended. No snippets, no pseudo-code, no isolated example. A person with modest or no coding experience needs complete executable as-is code and should not need to struggle with incomplete code that doesn't compile or execute. Same for the chapters devoted to acquiring and setting up hardware and software, these chapters give inclusive descriptions. Again, the neophyte should not need to struggle with grossly incomplete instructions and no hints to fill in the gaps.

Here is a story that contains computer code, even though computers and coding are not introduced in any detail until Chapter 7. The purpose of this story is to give some flavor to whet our appetite for the rest of this book, and provide a feeble rational for learning something about computer usage. The story is about Dick, Jane, and Spot, who are generic names that represent individuals, collections of individuals, entities, and anything else needed to move the story on. They live in a country very far away, and, their society has advanced to a level where national finances have become quite simple, apparently. The three characters are from the distant past, because, usually the past is a good predictor of the future. As citizens of the far away country, Dick, Jane, and Spot all pay taxes, and, do so by each independently and privately each year arrive at an estimate of their own personal fair share of the tax burden, expressed as a number between 1 and 998, representing their tax due as tenths of a percent of their personal total cash flow for that year, where the definition of cash flow is extended to include transactions, income, purchases, payments for services, barters, rents, gifts, bit-coin, property, interest, exchanges, gains, value fluctuations, compensation, trades, and all the various means to hide or avoid counting or avoid recognizing cash and other equivalents of transferable wealth. Note that only the percentage of personal wealth fluctuation is input data, not any dollar amount or equivalent. On the revenue sharing side, such as categorical grants, block grants, services, and all other means all which funnel money or some equivalent of wealth back into society, are determined in part by the taxes paid by each individual or entity of fair share, and, it is understood that sometimes those who pay more get more. All this wealth shuffling by the government is opaque to Dick and Jane. Spot developed and manages the operations of the financial software to do all this shuffling and more. When rolled out, the software was well received by the public and various experts, and respected appointed committees all claimed that the user interface was easy to use and wonderful and the application performed exactly as intended. Dick and Jane have a conversation about two years later. Both share that they had put in their good faith estimated fair share, but got back less benefits and services, and others got more than Dick and Jane knew was appropriate. When Dick and Jane spoke with Spot about their concerns, Spot said that the national government is elected and highly respected by all citizens, and Spot simply follows the laws, rules, and regulations established by the elected representatives and implemented by the executives. Spot further mentions that all the citizens are reasonably financially secure and generally happy, and only a very few are in poverty but taken care of with a social safety net. Furthermore the majority of the citizens approve of the way taxes and benefits and services are handled, and approve of Spot and the job Spot does. To wrap up this story, the following mimics the central part of the computer program that works behind Spot's user interface and wealth shuffling duties. In this code, simulated input, which is the fraction of total individual annual wealth fluctuation, expressed as tenths of percent, is generated automatically for illustration purposes and covers all possible tax due inputs inclusively. The output for each input determines where the corresponding tax collected is deposited. Taxes and revenue sharing works as follows: the computer input is an individual citizen's estimate of fair share. The code below shows how this input produces output. For all citizens one at a time, if the computer gives no output, or the output is a number less than or equal to 1000, then the collected taxes go into a government treasury account, which Spot distributes as bene-

CHAPTER 1. INTRODUCTION 7

fits and services and other responsibilities best addressed by government. If the output is greater than 1000 then Spot determines what account the corresponding collected taxes are deposited and how they are used. Spot's code:

```fortran
! this file name:   C_1_teneightnine.f90
! dependencies:  none
! to compile and execute use:
!              python C_1_doit.py
! this example code illustrates:
!              the heart of a story
  program teneightnine
!   many easy to find descriptions
!   of this on the Internet
   character*3 cnum
   do i = 0, 9
     do j = 0, 9
       do k = 0, 9
         if (i.NE.k) then
           ir = k
           jr = j
           kr = i
           num = i * 100 + j * 10 + k
           numrev = ir * 100 + jr * 10 + kr
           numdif = ABS(num - numrev)
           write(cnum,'(i3)')numdif
           if(cnum(1:1).EQ.' ')cnum(1:1) = '0'
           if(cnum(2:2).EQ.' ')cnum(2:2) = '0'
           read(cnum,'(3i1)')ii,jj,kk
           iir = kk
           jjr = jj
           kkr = ii
           numdifrev = iir * 100 + jjr * 10 + kkr
           new = numdif + numdifrev
           write(*,*)num,new
         end if
       end do
     end do
   end do
   end
```

How Spot compiles and executes the code:

```
# this file name:   C_1_doit.py
# dependencies:   C_1_teneightnine.f90
# to execute:   python C_1_doit.py
# this example code illustrates:
#    compiling and executing a Fortran program
import os
os.system('gfortran -o D_1_run.out \
          C_1_teneightnine.f90')
os.system('./D_1_run.out > E_1_see.txt')
```

In this story, those citizens who are computer literate, more or less, will more likely to be able to exert some influence over how society and its governance functions. Important to note is this story makes no value judgments regarding the actions of the characters, and each reader decides which characters are sympathetic and which are not. The computer and software make no moral judgments, only the people who cause the software to be developed and the people who use the software have the ability to add morality into the process.

This book is about thinking, likely many will say it is a how-to book for computer programming, with examples from probability, and abundant ramblings about water quality. However, thinking is the intended focus. This book takes the reader on a journey, the scenery and things happening along the way are activities that illustrate computer technology impacts on humanity, specifically thinking.

Starting this journey, consider that humanity has gone through two enormous advancements in thinking. The first is talking, verbal communication between two or more people. Talking took thinking from an activity isolated inside an individual's mind and expanded it to include activity shared among many individuals. Verbal communication is one major difference between humans and other living beings.

The second enormous advancement is writing, printed communication. Writing took thinking from an immediate localized activity within groups of people short distances from each other and expanded it over large distances in time and space. Writing also changed the way we think.

CHAPTER 1. INTRODUCTION

Preservation of ideas with writing no longer solely relies on memory and recitation.

Humanity is now firmly in the beginning phases of the third enormous advancement in thinking, which is computers and everything enabled by computer technology. Whereas written communication preserves and transports ideas through space and time, the written ideas are fixed by the process of printing symbols on media. Computer technology allows ideas and thoughts and other information to modify itself. In verbal and written communication, the modification of ideas still reside in the human mind that hears, speaks, reads, and writes the ideas. With computer technology, the modifications may occur independently of the human mind. Since the modification process is created by the human mind and implemented by human activity, what happens with computer technology is an extension, or an enlargement, or augmentation of human thinking. This computer aided extended thinking is a recent phenomenon and has been available for a short time compared to all of human history, and, the future promises to be different and amazing and beyond what a reasonably cautious person would wish to guess.

Thinking about, something. To make this book useful, it needs to illustrate the ideas with a concrete example. An over-arching idea selected here is assessment, which is abstract, and is not measurable with the usual physical devices known to science and engineering, such as a ruler or a clock or a thermometer. Also, assessment follows no rules such as addition and subtraction of numbers. Assessment is a process that happens inside the human mind and the human heart. Dogs likely perform assessments also, but communication between dogs and humans is not the same as verbal and written communication between humans and therefore humans and dogs can not have a discussion with each other on the subject of assessment. Assessment is influenced by culture, individual likes and dislikes, the business of people interacting, and, the ideas of what a desirable society is and how it works. So, assessment is a challenge to think about, and a challenge for computer technology to augment, or enhance, or enlarge the human thinking process.

Next select quality as the subject of assessment. Quality by itself is a challenging idea to use as an example of thinking. Like assessment, quality is an abstract idea, unmeasurable by physical devices, follows no mathematical or logical rules, follows no natural rules, such as gravity or light traveling in straight lines. As with assessment, quality is a challenge to think about, and a challenge to use computer technology to augment our thinking.

Finally pick something observable and measurable so we can think about quality assessment with a real world example. Select water, because it is ubiquitous, intuitive, very familiar, and is part of our every day lives. Water is sufficiently mundane so that strong emotional bias will less likely corrupt the attempt to think about quality and assessment, which can apply to anything, ranging from mundane to rather contentious subjects.

Separately, think about exclusive privileged access which is another theme in this book. Two historic extreme but illustrative examples may be found in one type of human society, hunter-gathers. First, hunter-gathers in a setting that sparely and randomly provides the necessities of life results in scattered nomadic societies roaming an endless resource deprived environment, mostly but not always peacefully with other groups living in the same environment and dependent on the same weather patterns. Second, when a place where people live provides a localized, constant, and reliable source of the necessities of life, society settles down and exploits the adequate and predictable resources, and, at some point in population growth, that society becomes possessive and aggressive about these resources. This is human nature, and evolves into exclusive privileged access.

Another historic example is knowledge, particularly knowledge as a resource that may be captured for the benefit of a few. In ancient China, so this story goes, the Emperor had a seismic detector that could indicate the direction of a strong earth quake that had occurred. The Emperor with this knowledge immediately dispatched large amounts of supplies and aid in that direction, and it could take days to reach the earth quake stricken region. To the residents of the region, this quick response was like magic, which the Emperor surely reinforced to gain their loyalty. The Emperor had exclusive privileged access to knowledge, and our society today has examples of privileged exclusive access, one example today is called intellectual property. This tradition is so ingrained in our society that many people consider it required, or absolutely necessary, or so normal that it is never questioned. Is the Emperor's privileged exclusive access to knowledge good or bad? Earthquake victims are helped as quickly as possible so it is good. The Emperor has absolute power and control over the subjects including the occasional earthquake victims, so depending on how the power and control is used, it may be good or bad.

Today, a reader of this book, who is not well situated, trying to follow along this case study, will not gain full benefit where access to computers and computer literacy are dominated by privileged exclusive access. Computer access and computer literacy are now following the same path that access to writing and reading took when they were adopted by humanity. Access to writing and reading, called literacy, had its ups and downs in history, and still has today. An intended consequence of this book is to increase the fraction of the population that has access to computer literacy.

CHAPTER 1. INTRODUCTION

Computer literacy has many facets. One is the computer replaces paper for information dissemination and preservation purposes. An example is accounting, and bookkeepers with software and a computer can be more productive and accurate than a bookkeeper with only pencil and paper. The bookkeeper's task is an example of enhanced thinking, aided by computers, aided by writing, compared to unaided mental computation and memory. However, whether the books are kept by mental arithmetic, or by writing columns and rows of numbers on a bookkeeper's page, or by entering or importing data into a bookkeeper's spread sheet, exactly the same task is performed, that is, adding and subtracting and keeping track of numbers.

Another path of computer literacy is using the ability to instruct computers to think along with humans thinking, in a manner that the interaction enhances human thinking ability. This is the primary focus here. In this book, with occasional philosophical tangents, we focus on thinking concretely about water quality assessment, using inexpensive available computer technology to enhance our thinking. And, for those with limited or no access for what ever reason, circumventing privileged exclusive access to computers and computer literacy.

Elaborating further, or belaboring about computer augmented thinking, consider arithmetic done in our minds only and done with the aid of pencil and paper, and how our exposure to arithmetic influences and augments our thinking. The Romans expressed numbers as Roman Numerals, an example is IX for what we call today the number nine. The Islamic world and other societies in Asia close to the time of the Romans expressed numbers as Arabic Numerals, which is how we express numbers today. The difference becomes obvious when considering long division. With Roman Numerals, long division or its equivalent is cumbersome and not used today, while long division today is known and understood by third grade or earlier. Long division of Arabic Numerals is an excellent example of how reading and writing augments human thinking, especially when the numbers are larger than we can keep track of in our heads. With the skill of long division, our thinking about numbers and doing things with numbers is significantly augmented, and we can think constructively and creatively about things that would be nearly impossible without long division and other manipulations of numbers. The benefits of thinking with the aid of pencil and paper are so obvious that our society has long encouraged these skills by having reading, writing, and arithmetic as the basis of our public education system.

There are two steps in this book, the first is access to computers and software, the second is using computers and software. The intent here is enabling computer literacy for a much larger percentage of the population than exists currently, which requires accessibility, which leads to general constraints that include frugality and simplicity. This is possible now that computers are commodities, readily available to everyone, which may be a goal of literacy. However, being such a new phenomenon, society has not yet settled on an established definition of computer literacy. Complicating the situation, computer technology usage changes frequently and dramatically, making opportunities for computer literacy appear and then become irrelevant haphazardly. Efforts to promote computer literacy are easily found, since there is a societal need for coders and other computer technologists. This book is one small additional effort, which may become a dead end, or a non-productive tangent, or spark an idea that offers an unimagined solution to an intractable but important problem.

Chapter 2

Review Precedents

At the beginning of our case study about water quality assessment, a review of how water quality assessment is done today is useful. The understandings and insights will guide our thoughts. Here, we make no analysis nor judgment about the merits, we concentrate only on the process and the basis of the process.

Many organizations world wide engage in water quality assessment, and in the United States includes industrial trade organizations, agricultural trade organizations, development and planning companies, environmental advocates, Chamber of Commerce, United States Environmental Protection Agency, United States Geological Survey, Department of Agriculture, and corresponding state and local governments. What and how these entities approach water quality assessment varies, and in the USA most efforts revolve around the rules and regulations promulgated by the EPA. So, EPA guidance, rules, and regulations are an excellent source of information about the process and the rational for the process, all of which may be found on the Internet.

Within the EPA and similar organizations world wide, water quality issues are divided into several distinct program areas, which include water quality standards, water quality monitoring, water quality reporting, water quality enforcement, or the equivalents by different names. Conceptually the process starts with standards, since without knowing what is good water quality and bad water quality, it is difficult to find an assessment process that makes sense. Examining standards, it quickly becomes apparent that different types of water bodies exist, and more important, these different types of water bodies are used for different purposes. Examples include drinking water, irrigation water, industrial process water, recreation such as swimming and boating and just enjoying the view, pristine waters, waters that support barge and ship transportation, fishing both recreational and commercial. Each of these different uses for water bodies get their own set of standards which describe what is good and what is bad water quality. These uses have a further differentiations, such as season, where in some parts of the country the water body is frozen in winter, while in other parts the water body is dry in the heat of the summer. As a result of differentiations of water bodies, each water body is given its own standards and one or more beneficial use designations.

After the standards and beneficial designated uses are established, the next task, conceptually, is monitoring, which includes taking and analyzing samples or measuring directly various properties and constituents of the water in a water body. Examples include temperature, dissolved oxygen, pH, arsenic, lead, nitrates, phosphates, E. coli, fecal coliform, suspended solids, turbidity, number of fish, number of fish species, algae, trash, garbage. These observations are the evidence that lead to a decision about the status of the water quality.

As the evidence accumulates, it is obvious that some consistent way to collect, preserve, and disseminate the information is needed.

Then, the evidence is examined, compared with the applicable standards for the various designated beneficial uses, and the result is a decision whether a particular water body attains standards or is impaired for a particular designated beneficial use, and this process is what most people call assessment. It's not over yet, of course, the assessment results need to be reported.

The intent is that the assessment reports will be studied carefully, and society will select options, make choices, all for the purpose of meeting and preserving the beneficial designated uses of each water body. Along the way, in our adversarial society, differences in opinion of various parties for which water is important are taken to court, and our judicial system decides who is right and who is wrong as far as the law is concerned. This leads to legislation and administration, where laws, rules, and regulations are revised, bringing politics into process. The courts and the legislatures, in different capacities, are the beginning and the final resolvers of the process that is called water quality assessment.

Chapter 3

Start Thinking

Start thinking about water quality assessment. Is there a different way to describe the essence of the process, but not be distracted by the presence of a bureaucracy, and focus on what assessment is rather than the business of assessment performed by some organization. Could we find a different way to think about the process and the basis of the process that is more unified, or, generalized, or attempting to understand as a whole rather than many unique special case parts.

Use computers is the direction we will take, since the real focus here is thinking with assistance of computers. Computers already play a ubiquitous role in water quality assessment, where computer technology is use to gather, organize, and disseminate the information generated by the monitoring process. In the United States examples include the National Water Quality Monitoring Council (NWQMC), the Water Quality Portal (WQP), which integrate publicly available water-quality data, through use of the Water Quality eXchange (WQX), from the USGS National Water Information System (NWIS), EPA Storage and Retrieval Data Warehouse (STORET), and USDA ARS Sustaining the Earth's Watersheds, Agricultural Research Data System (STEWARDS).

Exploring existing concrete details of the existing water quality assessment process, in the United States, there are published water monitoring program guidelines, such as "Elements of a State Water Monitoring and Assessment Program" which recommends ten basic elements of a holistic, comprehensive monitoring program that serves all water quality management needs and addresses all water body types. It describes a process in which States develop a monitoring program strategy to implement these basic components. There are published reporting guidelines, such as "Integrated Reporting (IR) Guidelines under Sections 303(d), 305(b), and 314 of the Clean Water Act", which are for states, territories, authorized tribes, and interstate commissions that prepare and submit Section 305(b) reports. The guidelines outline development of biennial Integrated Reports (IR) that support EPA's strategy for achieving a broad-scale, national inventory of water quality conditions and include listings of impaired waters under Section 303(d) of the CWA. Previous editions of these guidelines are also available on the Internet, along with "Consolidated Assessment and Listing Methodology", a website that describes the background, purpose and process for developing a "Consolidated Assessment and Listing Methodology (CALM)" that streamlines reporting requirements under Sections 305(b) and 303(d) of the Clean Water Act.

All this with computer technology, including the Internet, make it possible to organize and manage amounts of data that would be costly to manage in a system based on paper. On the surface, all this is the business of assessment. Underneath are the concepts that we're interested in here.

Chapter 4

Begin Understanding

Start with the following presumptive given, that the water quality assessment process takes observational water quality data and compares that data with the relevant water quality standards, and arrives at a decision which is binary, either the water body is impaired or the water body meets the applicable water quality standards. Include a second presumptive given, that the observational water quality data is a disparate collection of all sorts of information. The only generalization is that any one datum could be almost anything, such as a number, or a descriptive word, any number, any word. Include a third presumptive given, that an applicable water quality standard could be expressed in any way. It could be a paragraph, it could be a number, it could be a word. How can a standard and an observation, each of which is almost any type of information, be handled so that a computer could come up with a binary decision, impaired or not? And arrive at something in a way that makes sense to many people, most politicians, and agrees with the consensus conclusions of a large number of experts hired by all the adversarial stakeholders who are all looking at the same observational data and standards.

Computers can do many things, including handling numbers, so convert all the observations and all the standards into numbers and a computer can do it.

What are these numbers going to look like? How will they behave? How will observations and standards be converted to numbers? Most importantly, how will the politics of the process be addressed, because assessment involves human emotions and feelings, and these are far removed from anything that resembles numbers or could be called manageable in a computerized manner.

A problem is that assessment does not follow any rules of mathematics or logic or science or any voice of authority, and computers only follow rules of mathematics and logic. Yes there are sophisticated computer applications such as voice recognition, but deep inside these sophisticated applications are elementary processes using mathematics and logic.

Most people will be happy if a process is found that keeps the true final assessment decision in the control of the people. Enough people will be happy if the process happens on a level playing field. Some people will never be happy regardless of what happens. And some people will be unhappy enough to disrupt, discourage, destroy, discredit, disallow anything to happen, and start all over, which is exclusive privileged denial-of-access. Taking this further, consider the zero-sum game between the cat and the mouse. If the cat wins, the mouse looses, and vise versa. If the mouse refuses to play the game by not participating, by staying completely still and quiet, denying the cat the access to play the game, sooner or later the cat loses interest, forgets the game, or something, and walks away, and the mouse wins. The lesson here for assessment when computers follow only mathematics and logic is to arrange the assessment process so that politics and emotion and beliefs have the final say when push comes to shove. In addition, when the do-nothing strategy appears, it would be good if the assessment game is easily portable, inexpensive to duplicate, quick to set up, because then the other player has the choice to continue playing the assessment game or if possible take the game elsewhere and continue playing without the do-nothing player. That's what the cat does and the cat survives, and so does the quiet mouse.

The plan is to think in terms of numbers, which computers can handle. Likelihood may be expressed as a number. We need a procedure that will take observational data and take the applicable standard, and convert them each into a likelihood. Once we have a number for likelihood, we need a procedure that will allow observational data gathered over a period of time to be combined together to give the best possible estimate of the true likelihood that the water body is impaired or not. Let's see where likelihood leads.

Under well defined situations, the best possible estimate from data expressed as a likelihood can be found by using an equation called Bayes Rule or Bayes Theorem, where the likelihood is interpreted from the point of view of Bayes Rule. That is, if the numbers work properly when Bayes Rule is applied, then the result is the best possible estimate because it has been proven that Bayes Rule gives the best possible estimate. In other words, no other method can do better, another method may do the same, or worst, but not better. The reader is referred elsewhere to find a proof of this claim. Staying with more concrete ideas instead of abstract mathematical proofs, we now enter an expedient world of truth and understanding,

CHAPTER 4. BEGIN UNDERSTANDING 19

what does "proven" really mean? How is all this "understood"? How can a non-mathematician achieve truth and understanding equivalent in practical situations to the mathematician's formal proof?

Before proceeding into many bewildering details, the comforting words are that all of the following, which means the whole rest of this book, is not new. Every detail has been explored, studied, examined thoroughly in the past and may be found in many books and on the Internet. That means that everything is understandable if sufficient effort is applied. Whether what follows is perceived as easy or hard depends on the reader's past experience with and exposure to the details explored here.

"Understanding" is easier than "truth" which is subjective and emotional. Understanding is practical and useful immediately, and can lead to truth, while truth is something that may or may not be achieved sometime in the future, and may even be impossible to achieve. This distinction is particularly significant for quality assessment and binary decisions, where understanding how the quality assessment is conducted and understanding how the binary decision is obtained is more important than some abstract truth of the assessment process and resulting binary decision. For our use of Bayes Theorem, truth is that we understand how it works, and assure ourselves that it works consistently, reliably, and exactly as we understand it to work in our narrow scope of water quality assessment.

The numbers that work properly using Bayes Rule follow the axioms of probability. Now something can be said about truth, here, the axioms of probability are taken as truth. Or if you wish, follow the skeptic who may concede only that we assume the axioms are true and see where that leads to.

Water quality assessment protocols and procedures allow people to arrive at a conclusion, either the particular water body being assessed is impaired or meets applicable standards. The presumption by most is that a truth is obtained, either it is true that the water body is impaired or it is false that the water body is impaired. This language can not be taken lightly. Consider logic, and proofs based on logic. For example, a statement is made, and we want to determine if the statement is true or false. A straight forward procedure is to look for evidence that the statement is true. If we find evidence, that is not sufficient to say we have proved that the statement is true unless we examine all the evidence possible. If we find one piece of evidence that contradicts the statement then we say the statement is false. To prove a statement is true may be impossible if the possible evidence is infinite. To prove a statement is false needs only one piece of evidence.

From this point forward, no more philosophy about abstract ideas. Now we focus on the real world, where observational data is combined with applicable standards to give an assessment decision. This is where the real augmented thinking begins, resulting in understanding, more or less, what likelihood is, how probability works, what is Bayes Rule, and a few other things such as how to handle imperfect data and how to incorporate emotions and politics into the assessment process.

Chapter 5

Probability and Bayes Rule

What follows is a routine explanation of Bayes Rule, found in many textbooks and the Internet, followed by a rewording applicable to water quality assessment. Computer programs will be in later chapters, where the purpose is to give the reader opportunities to think along with a computer thinking. This chapter is an opportunity to do classical thinking, following chapters use augmented thinking by writing computer programs that eventually do water quality assessment, convincing ourself that the computer programs work correctly, and gaining understanding of the ideas expressed in the computer programs. Gaining understanding has limitations, the human mind is too limited to understanding vast complicated ideas, but, if the vast idea is represented by what the computer program does, then by putting together and understanding all the little parts that make up the computer program, more or less understanding is gained. More important, when finished the human mind has succeeded in taking a vast complicated concept and getting it to work, and produces evidence that it indeed works as intended.

What follows here is the most difficult concept in this whole book, derive an alternate form of Bayes Theorem:

Start by accepting that the multiplication rules for conditional probabilities are true, correct, and work properly for us, and write using probability symbolic representation as found in books and the Internet the following probability expression,

P(A | B) * P(B) = P(B | A) * P(A)

which reading left to right means: "the probability of A happening given that B happened, symbolized as P(A | B),

times, symbolized as *,

the probability of B happening, P(B),

equals, =,
the probability of B happening given that A happened, P(B | A),
times, *,
the probability of A happening, P(A)".
Next,
require that P(A) is neither zero nor one,
which means that A is neither impossible nor certain,
and require that P(B) is not zero,
which means that B is not impossible,
and then,
rewrite the above using algebra to get,
P(A | B) = (P(B | A) * P(A)) / P(B)
which means: "the probability of A happening given that B happened equals
the quantity (
the probability of B happening given that A happened
times
the probability of A happening
)
divided by
the probability of B happening".

An interlude is needed here. For most of the intended audience for this book, the above passage is nearly incomprehensible, at the very least the reader is left feeling uneasy about what is suppose to be gotten from this. The above is a result of someone in the past doing classical thinking. The technique of classical thinking is ponder an idea or something that is read, and ponder it some more, and ponder until what was written is understood, no matter how long it takes. Pencil and paper are allowed for thinking assistance. This is how classical thinking is done, there is no other way known. Classical thinking is sufficiently difficult that when a good idea appears and is published, a person gets their name attached to the idea. So, a long time ago a person named Thomas Bayes pondered about this and where this may lead.

On a fresh less daunting track, think about numbers. Think of what happens when things are done to numbers, such as addition and subtraction. There likely is no problem with understanding 1 + 2 = 3, and likely no questioning the truth of this. Same with understanding 1 + 3 = 5, and likely no questioning this is not true. However, not true to what? The simple answer is not true to how numbers are understood. Otherwise 1 + 3 = 5 looks like a perfectly reasonable way to write a sequence of characters, because it can be seen and we can read the sequence and every character in the sequence makes sense and is understood, separately.

CHAPTER 5. PROBABILITY AND BAYES RULE

Continuing on this track, start thinking about what happens when things are done to things happening to numbers. Cut a sliver off a long edge of a page of quadrille paper, giving a sort of a ruler. Using this ruler, the distance between two dots, in quadrille line separation distance units, may be measured. Put one dot on the page remaining, on the intersection of two lines, near a corner of the page of quadrille paper.

Depending on the size of the little squares printed on the paper, from the dot just placed in the corner, count over along the longer side of the page, some multiple of the number 4, such as four times five which is twenty. Put a dot on the intersection of the lines at that point, say 20 squares along the long edge of the page between the dots. Now going the direction of the short edge of the page, count squares until reaching the same multiple of the number 3, following the above, three times five which is fifteen, and put a dot on that intersection. Take the ruler and measure the number of square-lengths between the second and third dots. The result will be the same multiple times 5, which here is five times five which is twenty five.

Now think about the numbers, 4, 3, and 5. Is there a relationship? Most middle schoolers would credit the Oz movie for knowing the relationship, and Pythagoras is credited with the first proof of this ancient relational concept. Did he write about or talk about what he was thinking about and the understanding he gained from that thinking? More than 100 years later Socrates bemoans that writing spoils the mind. Advances in thinking evolve slowly, even in the minds of the greatest thinkers of the time.

Back to the dots on the graph paper, which may be represented by numbers. In the example here, the numbers are the number of little squares counted on the graph paper. The importance of this is that a dot can be represented by numbers. Because this is possible, computer technology can do all sorts of things with dots, not with the dots themselves, but with the numbers which represent the dots.

A square root is something happening to a number. In the example here, square roots and their inverse, the square, are things that happen when the distance between dots is calculated using Pythagoras's triangles, that is, get another number after doing things to other numbers, and all the numbers have to do with representing dots on a piece of paper, and numbers representing things happening to dots.

Getting back to thinking about what happens when things are done to things happening to numbers. In this example, lift the paper and turn it around arbitrarily and put it down, or put it on a wall, or on the ceiling. Then using the ruler, measure the distance between the second and third dots. The result is the same as before. What's happening is the dots are moved through time and space, and the distance remeasured. Moving

CHAPTER 5. PROBABILITY AND BAYES RULE

through time and space is things happening to the two dots, and happening to the measuring the distance between the two dots, which are represented in computer technology as things happening to numbers. So, dots moving through time and space become something that computer technology can handle. Video games are one application of this, and all the excitement is based on some coders conceptualizing with numbers things happening to dots in space and time.

Analogous to dots moving through time and space, if water quality assessment concepts can be represented by numbers, then computer technology can handle that. Here, water quality assessment concepts will be represented by probability. Probability may be expressed as a number. The door is open for us to use computer technology to augment thinking about water quality assessment.

Rewording Bayes Theorem using water quality assessment verbiage:
"The probability of the segment A being impaired given the evidence B

equals
the quantity (
the probability of the evidence B given the segment A is impaired
times
the probability the segment A is impaired
)
divided by
the probability of the evidence B".

Back to the symbols that express ideas of probability, another form of Bayes Theorem may be derived from the above. Accept as true and start with a known relationship that occurs when the two happenings, A and B, are independent,

$P(B) = P(B \mid A) * P(A) + P(B \mid .NOT.A) * P(.NOT.A)$

Substitute this into the multiplication rule for probability, above, which gives,

$P(A \mid B) = (P(B \mid A) * P(A)) / (P(B \mid A) * P(A) + P(B \mid .NOT.A) * P(.NOT.A))$

which is the alternative form of Bayes Theorem used here.

Further elaboration, accept as true that,
$P(B \mid .NOT.A) = 1.0 - P(B \mid A)$
and
$P(.NOT.A) = 1.0 - P(A)$.

Rewording the above alternative form of Bayes Theorem, for water quality assessment:

The probability of impairment A given the evidence B, written symbolically is, $P(A \mid B)$,

CHAPTER 5. PROBABILITY AND BAYES RULE

and, the numerical value of this probability is the result of the calculation located on the right side of the equal sign, which consists of a numerator and a denominator.

The numerator is,

the probability of evidence B given impairment A, written symbolically is, P(B | A),

and which, as described later, is obtained by using a translator to convert data to probability,

times

the prior probability of impairment A, written symbolically is, P(A).

The denominator is

the numerator, above,

plus the quantity (

the probability of the evidence B given non-impairment .NOT. A, written symbolically, P(B | .NOT. A),

rewritten as (1.0 - the probability of the evidence B given impairment A), or, (1.0 - P(B | A))

times

the probability of non-impairment .NOT.A,

rewritten as (1.0 - the probability of impairment A, written, P(A)), or, (1.0 - P(A)).

The final result is the numerator divided by the denominator,

where all quantities have readily available numerical values,

which include:

the prior probability of impairment A, written, P(A),

the probability of the current evidence B given impairment A which is obtained from a translator that converts data to a probability,

and a result is calculated, which is the probability of impairment A given evidence B,

which then becomes the prior for the next calculation with new evidence.

There are likely several places in the above that deserve closer scrutiny, one is: "the evidence, B, is required to be independent of the impairment, A." At this point, we are asked to accept the relationship as being true and continue the derivation. We could have stopped at that point and ask, is the impairment really independent of the evidence, Doesn't the evidence exist because the impairment exists? Doesn't that make them dependent?

We are now thinking, and pondering, and mulling. Situations such as this is where truth and understanding can be determined and achieved, or not.

As often happens, our vocabulary and how we understand what words mean is the difference between understanding and confusion. In this case, perhaps we should not use the words evidence and impairment, but rather use the words observation and assessment, which most people will agree are independent events, where evidence and impairment are outcomes of events. Back to the simple example, a coin flip is an event, heads and tails are outcomes of an event. Flipping two coins is two independent events, which yield two independent outcomes.

Most of this chapter illustrates, perhaps, why arithmetic and mathematics are usually approached with symbols rather than words. All this will be revisited again in Chapter 16 at which time these same concepts should be easier to navigate.

Chapter 6

Making Choices

Before moving on to actually using computers to augment our thinking, access to a computer and appropriate software are needed. Exclusive privileged access is fine, but access for everyone and anyone is preferred, and demanded if computers in our lives follow a similar path as talking and writing followed.

First, an assumption about the capabilities and resources that a reader starts with, because the rest of this book is based on that assumption. Assume the reader has no experience with the Linux operating system, and of course does not possess nor have ready access to a computer with a Linux operating system. It is assumed that the reader has an education some where in the range of High School and an advanced degree, and is not completely ignorant of computer technology in our modern society, examples but not prerequisites include word processors, spread sheets, Wifi, Internet surfing, smart phone, ripping a DVD, and hopefully some computer programming experience. Mathematics beyond algebra is not needed here.

Now, having no computer, what computer to use? What programming language to use? Two good questions. Ask around, surf the Internet, and there are 10,000 ideas about what's best and what's worst. And, it boils down to what do you want to do? Construct a Web page? Then use HTML. Edit photographs? Then forget programming and use one of the many photo editing software applications, unless image manipulation is really what you want to do.

What programming language should be used to write a computer application that augments and enhances human thinking? There are many programming languages to choose from, and many are good and usable.

To choose a computer and a computer language, examine the desired

CHAPTER 6. MAKING CHOICES 28

outcomes and the constraints. Frugality is a goal and may be a constraint by necessity, along with simplicity, transparency, ease of use, understandability. To achieve frugality, let us purchase a used or salvaged computer and hope everything else is free. Purchase something widely used, so there is a used market, and, something easy to have around, such as a PC laptop. Something easy to carry around and set up literally any place any time. Something cheap enough that if it breaks another may be bought cheaper than it would cost to fix it. Another source for a computer is a friend tossing out an older computer. Another direction is the Raspberry Pi. The cost of a working salvage laptop and a working Raspberry system are both far less than a new personal computer. The effort to get either up and running are about the same. For all the coding in this book, they are equivalent.

The laptop used for all the programming in this book was purchased without an operating system, at a salvage outlet for $20 with a three month guarantee. It may have knocks and scratches, things may be missing, but it is functional, not too old so it has adequate CPU speed and memory, and other necessities such as WiFi and a CD/DVD drive and USB sockets. Missing are the operating system and the battery, and it did not come with a power adapter. A power adapter was purchased separately and it runs fine even with no battery. For those who like to tinker with the parts, the Raspberry Pi works just as well though slower and takes more money, but it is much less portable, it's difficult to take it to the local library and set it up. Also, $20 is really cheap, and, it may be difficult for a reader to find a similar deal.

Next is software to put on the laptop, and the first software is the operating system. Now the questions are, what about simplicity? What about viruses and malware? What about support if we can't get something to work? How difficult will it be to recover from some disaster? What about Internet access? What about our productivity? Are the tasks we wish to do quick and easy? What about the learning curve? Are we productive quickly or will it take several months, or longer? Some of these question are hardware items. Regarding software, the short answer among many just as good short answers, is use a current distro of Linux, which costs nothing except the time and energy to find it, download it, and install it. If you have money to spend, go for what ever you wish and it will work too, just as good but a little bit different than and in the long run cost considerably more than the frugal simple choice for this book. In this book, we take the short answer, Linux, satisfying the goals of frugality and access.

Computer technology today, will be different tomorrow. The dominant computer corporations continuously bring out new, better products, and these companies and the public allow existing products to eventually

CHAPTER 6. MAKING CHOICES

fall by the wayside of disuse. And for good reason. The new products have more and better capabilities. The older products are about as useful as a buggy whip in today's age of automobiles. For frugality purposes, a recycled laptop is used in this book, but, it is not outdated. Most important, some version of Linux will run on it. Yes, a particular application may exceed the laptop's capability, and if that happens and the application is important enough to spend the money needed, then get a different computer that can to the job. If you can put Linux on it, it will work for everything here with no additional effort.

The particular version of Linux used here is called Crunchbang. Two other options are almost equivalent, which are Ubuntu and Raspbian. Typical of the technology today, Crunchbang is no longer actively supported, so, it has been cast on the trash heap. If that bothers you then use what adopted and resurrected Crunchbang, which is Bunsen Labs, or, select Ubuntu, or, Raspbian. All four are equivalent, that is, the computer code illustrations in this book work on all four. Yes, there are minor differences, and anyone who codes learns to live with that. As a side, Raspbian is available for installation on a PC, it is not limited to installation on a Raspberry Pi.

An Internet search will find a site quickly such as "LinuxFreedom.com" which archives older software and makes Crunchbang available for download, free. All the details later. How was it selected for this book? It was found while randomly looking for a small simple Linux operating system, and the looking stopped there, because there was no compelling reason to continue looking.

Documentation of the operating system is essential. Crunchbang, being defunct, has documentation created during it's short lifetime. Ubuntu and Raspbian and Bunsen Labs are currently supported, and all have excellent documentation on the Internet. The instructions for obtaining, installing, and using Crunchbang, Bunsen Labs, Ubuntu, and Raspbian are the same, that is, you can use documentation of one for another if you can get past the minor differences. This is possible because each operating system is derived from the same foundation, which is Debian, which is another version of Linux.

What about viruses and malware? This and several related risks need close consideration, it comes under the broader risk called disaster, everything is lost, for what ever reason, and, the disaster is unavoidable and unpredictable. The strategy here is be prepared to replace everything at any moment, and that includes the frugal salvage laptop, the Linux operating system, copies of what ever programming and gathered data that has been developed and regularly archived in at least triplicate, and even includes replacing the person who does all the work. Preparing for re-

CHAPTER 6. MAKING CHOICES

covery from disaster here is simple. By going through this book, you are qualified to become the replacement, same for anyone else. Prepared this way, when disaster occurs, or some persistent annoying problem crops up, replace what ever needs replacing. For a total replacement, the time required starts with one minute to put the replacement laptop on the table, plug it in, and get out the Crunchbang or your preferred "live-CD". Installing the Linux operating system on the laptop takes 15 to 30 minutes, more or less, depending. After checking to see that everything is working, about 5 minutes, pack up the laptop and head for the local public library, or where ever free WiFi access is available. After 5 minutes or less to set up and turn on the laptop, connect to the library public WiFi. Update, then install what's needed. Take another 3 minutes to make sure it works. Then go home again, and copy all your archived files onto the new laptop, and you are up and running, which will take anything from several seconds to the better part of an hour, depending on how much is being restored from your archive. Total clock time, depending on travel distances, less than half a day, more or less. Complete details for all this is later in this book, and starting from scratch is almost the same as a complete recovery from a disaster.

What if help is needed? When working all alone, there is only one good answer, search the Internet. When working inside some organization, find a knowledgeable person. However, a knowledgeable person always knows less than the Internet, and always has more bias than the Internet. Of course there are exceptions, and, a knowledgeable person gives a useful response much quicker than may be found on the Internet.

What programming language to use? This question comes with much preconceived bias. So, enter the world of computer languages, and find that several distinct approaches have evolved. Not surprising, considering the Tower of Babel. The motivation for different approaches is use, and, modern computer technology has blossomed with different uses. One use is number crunching, and an approach called procedural handles that quite well. A second use is establishing complex enterprises whose activities include transferring, translating, manipulating information on a large scale, and, another approach evolved to handle this, called object-oriented. There are other approaches, many others, but there is no established list of programming language types.

Our purpose here closely resembles number crunching. There are many computer languages that do a good job at number crunching, we'll consider three, Python, C, and Fortran, because these three have or had name recognition. All three are potentially good for our purposes.

Both Fortran and C are procedural computer languages. Both over time evolved to include the capabilities of object-oriented computer languages,

CHAPTER 6. MAKING CHOICES 31

Fortran has not changed its name, just changed the suffix, and C grew to include C++, which may be a new name or just a suffix.

Python is different. Both Fortran and C are old and established in their niche. Python is relatively new, and was developed purposefully to included several computer language approaches, procedural, functional, and object-oriented. Python also is evolving, and a suffix is added to the name.

The rational for using the procedural approach here, is that it produces computer code that is, in general, short and readily understandable, so therefore is the best for this book where the reader is presumed to be mostly inexperienced with computer programming, but not ignorant of other amazing products of computer technology and therefore sufficiently motivated to continue here.

If the code is kept simple, which is a goal, then it is almost but not quite trivial to translate code in one language to another, which makes computer language almost a non-issue. Just keep the code simple by using coding styles that are as elementary as practical, avoid cute and clever approaches, avoid obfuscation and other ways to hide what really is happening in the computer program. The code should reflect the programmer's ability to represent complex abstract concepts in simple efficient understandable transparent code, rather than reflect how clever the programmer is exploiting arcane and cryptic programming that demonstrates a complete and thorough knowledge of the compiler and the programming language capabilities and quirks. Any such faults are usually in the programmer's ego, not the programming language.

All this advice is for the solo coder, which is one extreme on a continuum that ranges to the other end, a coder among many coders in a large organization developing a very large enterprise application, where the advice is do what your supervisor says, even if that results in arcane and cryptic programming.

Chapter 7

The Details to Get Set Up from Scratch

Motivation in excess will be needed to complete this task when there is no prior experience and everything is new and bewildering. This task should be simple and easy, but is not. Privileged exclusive access and wrong assumptions about your knowledge of laptops and Linux will be thrown in your face. Not nice. This is the way the world has evolved, it is accidental and no one has cared enough to correct the situation, or, someone benefits from the way the world is, and prefers it does not change. After completing successfully a few times, this task becomes simple and easy and quick.

The task may be divided in two, 1) get a computer, and 2) get a copy of an operating system. This may be done in any order, but concurrently is best because a computer and an operating system are interdependent. Doing each task will shed understand about the other task. This is a tough learning experience, the most difficult task in the whole book, with false starts and disappointing outcomes. So after going through the process and succeeding, help someone else with the process when it is their turn.

Computer Acquisition Task: Frugality is an objective, so find a computer store or a salvage outlet or someone who sells used computers at a reasonable price for a laptop that may be old and looks used. A desktop works equally well, may be easier to find and may be cheaper, but is not portable so has other limitations. Select a laptop that has the following guaranteed working capabilities: 1) Wifi, 2) a hardware switch on the side of the laptop that turns Wifi access on and off, 3) USB ports, 4) optical drive able to read CDs and DVDs, and hopefully burn the same, 5) a hard-drive, 6) plug-in power supply, 7) a battery is nice, but not essential to work, 8) a screen, keyboard, and pointing-pad, of course, 9) no oper-

CHAPTER 7. THE DETAILS TO GET SET UP FROM SCRATCH 33

ating system, because installing Linux will destroy any operating system already installed and you've wasted a lot of money, 10) a guarantee for at least several months, to replace the laptop if it doesn't work, 11) ability to modify the boot sequence in BIOS. Look for a sticker on the laptop giving the version of Windows that was factory installed. If it is Vista or later, great. If it is older than XP, keep looking. Talk with the person selling the used computers about each item specifically, making sure all capabilities are present and working and guaranteed. Tell them you'll be installing Linux and if it doesn't work after a good faith effort, you'll return it on guarantee and walk out with another. If you get no satisfaction, try somewhere else where scoffs and smirks don't happen.

Operating System Task: If not already available, get access to a computer for about half a day total, which allows you to 1) have unfettered access to the Internet and the ability to download and save a large file, and 2) be able to burn the downloaded file onto a DVD, both as downloaded, and as a "bootable image" sometimes called a "Live-CD". Read the download and install information for Ubuntu, Bunsen Labs, and Raspbian on their website first, regardless, because the details are the same when you accommodate the minor differences. If it interests you, try finding documentation for CrunchBang on the Internet.

Find on the Internet a Website from which you may download an ISO-file that is a copy of the Crunchbang Version 11 Linux operating system, or at some future time, what ever is currently functionally equivalent to what Crunchbang is today. Some alternatives are Bunsen Labs, Ubuntu 14.0, and Raspbian. There will be several sites for Crunchbang, there will be several versions, and many of the websites will be troublesome one way or another. The website you end up using will be easy and straight forward, and no redirections to other Websites. X-out of any Website that throws fluff at you or tries to redirects you. For Ubuntu, use the Ubuntu website. For Raspbian use the Raspberry Pi website. For Bunsen Labs use their website. For either Crunchbang or Ubuntu or any other distro, download the version that is "32 bit" and "i386", which should work on any used computer you'd consider, and performs very well, as you will see. If you are at the library, downloading using ftp will take between less than one-half hour to more than four hours, depending on how the library has it's WiFi and Internet set up. If it takes longer than one hour, find another library, or something. When the download is complete, check that it is successfully saved in the computer you are borrowing, by checking the size of the saved file with what the Website says it is. The size for Crunchbang Version 11 will be more than 700 MB. A CD-R holds 700 MB, and a DVD-R holds 4.7 GB. Crunchbang is right on the edge between able to use a CD and needing to use a DVD. Ubuntu is larger and requires a DVD. A version of Raspbian is available that installs on a laptop, download it from the Raspberry Pi website.

If you don't succeed downloading, as described above, then on the Internet, use your favorite search engine and learn in more detail what an ISO file is, and, learn in more detail how to download and save an ISO file, and, how to burn that file onto a "bootable DVD". Trying to do it first with minimal instructions will help learning now. There will be different instructions for Windows, Linux, Apple-OS, Android, and others, and, different instruction for different releases of Windows, Linux, Apple-OS. Learn about taking a bootable DVD made from an ISO-file and installing the operating system that is contained on the bootable DVD into your newly purchased used barebones laptop. Barebones means it does not already have an operating system installed. If it did, you've paid about $100 more than an equivalent barebones laptop.

Then do the installation. Start with the laptop unplugged. With a bent-out paper clip, put the end into the tiny hole next to the button that opens the optical drive tray, and push. The tray will open. Put your "Live-CD" in the tray, make sure it spins properly, and close the tray. Plug in and push the power button so the laptop turns on. Immediately push the correct F-key that interrupts start-up and allows you to modify the boot sequence in BIOS. Different makes of laptops use different F-keys for this purposes. Do some Internet homework and find out how to modify the boot sequence on your particular recently purchased used laptop, or ask the person who sold it to you as you are buying it. Modify the boot sequence so the optical drive is tried first. Then exit BIOS and continue, which means do nothing and allow the laptop to boot. Follow the installation prompts that appear on the screen and choose the default option when given a choice, because at this time the options presented are not understandable. For Crunchbang, you will need to enter a Username and Password, one for the system, one for yourself. Keep the username and password very short and very simple, three lower-case letters is enough, for convenience use the same Username and Password for both the system and yourself, and remember them because you will need them later, which means every time you turn the laptop on. No one will be interested in breaking into your Linux computer, so security is a low priority. When installing the system, or any later time, you can't break anything accidentally, but you can mess things up. You can start over if you need to. You are likely to succeed the first try, on everything you do.

The operating system, Linux, is now installed, and you are finally comfortably at home with your new frugal laptop with Linux. Plug in the power supply, press the power-on-button and the laptop will turn on. Enter your Username and Password when prompt. Wait until the small icons

CHAPTER 7. THE DETAILS TO GET SET UP FROM SCRATCH

appear along the top-right the screen. Ubuntu and Raspbian look and behave differently. Learn by trial and error what happens when you place the cursor various places on the screen and press the right-click and left-click buttons on the touch-pad. A menu drop-down box will appear when the cursor is almost anywhere and right-click is pressed. Close to the top of the menu is Terminal, and the bottom is Exit. How to shut down gracefully? Get the menu, place the cursor over Exit, left-click, get another pop-up, place the cursor over Power Off, left-click, wait until the screen turns black, and unplug the power-supply. Again, Ubuntu and Raspbian are definitely different. In an emergency, hold down the power-on-button for several seconds, until the laptop turns off.

Information specific to your laptop about Wifi access to the Internet is found by placing the cursor over one of the little icons along the top-right edge of the screen. Simply letting the cursor sit on each icon will bring up a small box with a few words indicating a status or identification of each icon. In this manner, trial and error will shortly identify the "network connections" icon. With the cursor over an icon, left-click and right-click will bring up further information. For the connections icon, the information indicates whether or not the hardwired switch on the laptop is toggled to "allow connection" or "disconnected by hardware switch". When the hardware switch allows connections and the cursor sits on the connection icon, after a few seconds so your laptop has some time to search for available Wifi sites, left-click, and the information box will appear and display all the available Wifi-spots currently detected by the laptop. If you are in an urban area, there will likely be many spots listed, all locked. These spots belong to your neighbors. If you are in an unpopulated valley in the middle of no where, the laptop will find no Wifi sites.

With the newly installed Crunchbang, the web browser is out-of-date and no Fortran compiler is installed. Ubuntu and Raspbian downloads are current but also no Fortran. Everything else is fine and usable as is. So, back to the library or where ever you get access to the Internet. Plug in and turn on the laptop. Flip the Wifi hardware enable switch to "enable connections", and wait a few seconds. Left-click to get Wifi availability information. Find in the list the most likely spot hosted by the library as free public access to the Internet. Left-click on that line in the list, and either your laptop will connect or not, or a box will appear asking for a password. If you have problems, ask a librarian if Wifi is working, however, be prepared to get a shoulder shrug if you mention Linux. Many times the librarian needs to push a reset switch on the library's equipment, while all of the other library patrons with their laptops out are doing nothing because they are embarrassed about their ignorance and inexperience and are waiting for someone else to ask the librarian. Pack up and leave if Wifi is not working at this time, too often it takes too long for someone to fix the library's public Wifi if it is not working, after all, it's a library and libraries are all about books.

If everything goes well, your laptop now is connected to the library free Wifi. As described above, activate the menu box and left-click on Web Browser. If you are not connected, the Web Browser will tell you with "Server Not Found". If you are connected, you will see the default welcome screen for the Web Browser, or, you will see the library's welcome screen usually asking you to click indicating you promise to follow the library rules and do nothing bad. If this appears, you can not continue until you click "agree".

The Web Browser in Crunchbang is named Iceweasel, which is a version of Firefox, which you may have seen before, so no details are provided here about how to use it. But, as mentioned above, this just installed operating system is old and out-of-date, because all web browsers go out-of-date all the time as new Internet features are introduced to the world and new viruses and other bad things are born. So, the web browser needs replacing with the most recent version. The web browser comes already updated in Ubuntu and Raspbian, and are different than the web browser in Crunchbang.

To update the web browser, close Iceweasel if it is running, by a right-click on the little X in the upper-right corner of the Iceweasel window, and you are back to the dark gray minimal screen of Crunchbang, but your laptop is still connected to the library free public Wifi, which you may check using the small "connections" icon. Activate the menu box and left-click on Terminal. A medium sized box with black background will appear and you will see a flashing cursor immediately following three characters :~$ which mean something to people familiar with Linux but mean nothing to most people.

You are now in the control room of a vast and complicated world. The controls are primitive and like fossils show what life was like at an earlier stage of the evolution of computer technology. The controls are a keyboard and a screen.

You press a key and something happens out of sight and sound somewhere in the vast complicated world. With the usual luck and skill, characters appear on the screen. There must be movement of some kind, some where. Things can and are happening and you are in control. This is like Aladdin's lamp, you rub it and the genie appears, "I am your faithful and trusted slave and my only purpose is to make your wishes true. What are we going to think about today?" You now have that genie, your very own, and thinking together opens worlds to you unrealizable in the paper and pencil past.

CHAPTER 7. THE DETAILS TO GET SET UP FROM SCRATCH 37

With the keyboard, enter

sudo apt-get update

and press the Enter-key. Because you entered *sudo*, which means you need privilege to do things, you will be asked to enter the system password, do it. Sort of like exclusive privileged access under the excuse of preventing trouble caused by inexperienced people. Lots of characters come spilling on the screen. Wait until you see :~$ and the flashing cursor. Now your laptop knows about the most recent versions of various applications, one of which is Iceweasel. All this comes to your laptop from the Internet.

With the keyboard, enter

sudo apt-get remove iceweasel

and press the Enter-key. If all goes well, again lots of characters will spill on the screen and stop when you are asked permission to continue, because lots of bad things may happen if you really don't want to do this. Usually your response is an upper-case Y, and lots more characters spill on the screen, until it is finished and you see :~$ followed by the blinking white cursor. Iceweasel is gone.

With the keyboard, enter

sudo apt-get install iceweasel

and press the Enter-key. Characters spilling over the screen look familiar now, give keyboard response when requested, and wait for it to finish. Now you have the most recent version of Iceweasel installed, so you should have much fewer problems on the Internet than you would if you used the out-dated version that came with the operating system.

For Crunchbang, Ubuntu, and Raspbian, now install Fortran, in the Terminal with the keyboard, enter

sudo apt-get install gfortran

and press the Enter-key, and as above until finished. Now your laptop is ready to do any and all the examples in this book, and much more. Reflect a moment on Aladdin's genie, the tap of your finger now has great power, more than you realize. You may x-out of Terminal at this time.

Whenever you climb a tree, you should first decide how you are going to climb down, and what you'll do if a branch starts to crack. Computers are similar. To shut down, first, x-out of any open applications, which will appear as icons along the left-top edge of the screen. If you are sensitive to security risks, use the "connections" icon to "disconnect" from the library Wifi service, and, set the hardwire Wifi switch on the laptop to "hardwire disconnect". Now your laptop is safe and secure from any nasty bad guys trying to sneak into your computer, and more important, put your paranoia at ease. Even though you can relative easily and quickly and at no cost recover from any disaster, by simply reinstalling the operating system from scratch and reloading any archived data and information that you consciously and frequently backup, and, even though you rarely if ever press a sequence of keys giving away a credit card number or a social security number or a bank account number or anything useful to a nasty bad guy so you don't need to worry about that, cyber-threats are out there and are serious and require your diligence. Worst of all, these threats are bothersome, wasting your time and energy, and unfortunately, evolve so you do not know what to expect. If you are connected to the Internet, and, are unsure and frightened, hold down the power-on button on you laptop for a few seconds, until your computer shuts down. Then start it up again. Your time is better spent doing other things with your laptop than fighting obnoxious malware. If the worst somehow happens, recover from disaster by reinstalling everything. After it happens once, you will be a true believer in the frugal fiercely independent and self-sufficient approach to computers and software described here. Yes, unfortunately, it will happen sooner or later. Since you are a self-sufficient solo coder and not embedded in some organization which provides all computer protections needed, there is nothing you can do about that.

Next, how to write, compile, and execute computer programs. Python and C come with the operating system. Fortran which doesn't come with Crunchbang or Ubuntu or Raspbian, has been installed, above.

But first, we need to know how to get around on the computer, specifically for coding. Most if not all readers will already know, to get around, with the mouse move the cursor to an icon on the screen, then click on that icon. This is basically how Ubuntu and Raspbian work, and this is a big plus to select Ubuntu or Raspbian. Crunchbang is more elementary, some may call it primitive, the polite term is minimal. Regardless of the operating system being used, we are going nowhere except frustration if we don't know how to get around. This may be exceedingly elementary, but, it is absolutely vital to do actual coding on a real laptop.

The basic places are: 1. how to turn the laptop on, 2. how to turn the laptop off, 3. get to the "file system", 4. move around in the "file system" which means going to different "directories", 5. how to create and destroy "directories" also called "folders" in the file system, 6. how to see the names of the "files" in a "directory", 7. how to "delete" a "file", 8. how to "move" a "file", 9. how to "copy" and "paste" a "file", 10. how to recognize different types of "files" by looking at the file name, for example a text file name example is bigbook.txt and a PDF file name example is interestingbook.pdf, 11. how to "run" an executable file which in other operating systems are called "apps", 12. how to get to the "Terminal", 13. how to do all the above while in "Terminal", 14. how to start the "text editor" to write a new file, 15. how to open the "text editor" for a file that already exists. All this is not intuitive and very daunting for some one who has never seen this happen. Unfortunately, our society presumes every one has these abilities before entering first grade, or shortly after. So, don't be bashful and ask someone to show you, it will take about five minutes. Unfortunately, and purposefully, operating systems differ from each other and Linux is different than other operating systems, and sometimes equally frustrating, almost equivalent Linux flavors such as Crunchbang, Ubuntu, and Raspbian also have minor differences. As a result, all this is likely new for many readers. Since Linux gurus appear to be far and few between, the neophyte reader either must hack all this, or, look it up on the Internet, or, something. So if you are really starting close to scratch, be prepared to spend a few hours rather than a few minutes, depending on the extent of your previous experience and access to pertinent information.

The default text editor that comes with the Crunchbang operating system is named Geany. Ubuntu and Raspbian both have different default text editors. Geany is really a specialized text editor for coding, and is rather versatile as you will discover when you explore the options it offers. In Crunchbang, open it from the menu, under the name Text Editor. Enter the following with the keyboard:

i = 23

print(i),;print(" is a number")

with no leading spaces for each line. Save this by a left-click on File in the upper-left corner of the Geany window, then click Save As, then fill in the name you wish, but, the file name must end with .py or the computer won't recognize it as being a Python program file.

Open File Manager from the menu, and see that the file actually successfully was saved and named as you wished.

Next, from the menu, open Terminal. If you haven't created any new directories yet, nor moved into a different directory while doing the above, Terminal should have the file you just wrote and saved. You can check by entering ls, that is, lower-case "ell" "ess", using the keyboard and then press the Enter-key. The name of the file you just saved should appear on the screen.

Python does not use a compiler, whereas both C and Fortran use compilers. Python uses an interpreter which does essentially the same as a compiler but on-the-fly. Say for example, the file you just saved is named my23file.py. To run this Python program, in Terminal enter with the keyboard

python my23file.py

When you press the Enter-key, the program executes and the output appears on the screen. That's it, however, programs and output can be much more complex than this example.

An example for a C program follows, with the Text Editor enter

#include <stdio.h>
main()
{
int j;
j = 23;
printf(" %d is a number\n",j);
return 0;
}

where extra spaces in the lines are ignored by the compiler, so indentation only satisfies our aesthetics as to how a computer program should look. As before, save the file. As an example here, the saved filename is my23file.c. To compile, in Terminal as before, enter:

gcc my23file.c

When this line is executed by Terminal, it compiles the program and creates an executable file named a.out, and puts it into the same directory we already are in. To execute, still in Terminal, enter:

./a.out

When the Enter-key is pressed the compiled program is executed and the output appears on the screen, just as with the Python program, above.

Fortran is similar to C. With the Text Editor, enter

program my23
i = 23
write(,*)i,' is a number'*
end

Save it as, for example, my23file.f90. To compile, enter Terminal as above:

gfortran my23file.f90
and execute by entering
./a.out

and the output will appear on the screen. Notice that compiling and executing C and Fortran look rather similar, they both produce a file named a.out. This is because, according to the documentation, the gfortran compiler is a front-end, and converts the Fortran code into C code, which then is fed into the C compiler, which produces an executable file named a.out. The gyrations of executing by entering ./a.out is needed so that the computer knows where to look to find the executable file when the computer is instructed to run the executable file. This gyration is one of very many quirks and idiosyncrasies of Linux, and is a reason many people are more comfortable with clicking icons on a screen instead of entering commands in Terminal with a keyboard. The control room evolved before ease, familiarity, and comfort evolved in the computer technology world. To use another analogy, a bulldozer does not drive like a luxury sedan, nor a race car. The Terminal screen is not a video game nor virtual reality nor a smart phone screen.

To learn more and get more details about what is covered above, consult the many text books on Python, C, Fortran, and Linux, and consult the many tutorials and manuals located on the Internet for the same. It is presumed here that the reader will glance at some of these briefly, and then push on, because that's all you need to do, unless you need particular information about some tiny but significant aspect of coding. Then, it is easiest to just write a query in the input of your favorite Internet search engine, and many answers will appear for you to read. Most will beg the question, or inadvertently be misleading or otherwise useless, but a few will give you the answer you are looking for and in the words you understand. Asking the Internet often is quicker than consulting a tutorial or a reference manual, after you find a useful answer, and you try it, and it works. However, keep the manuals and tutorials handy. A last word, good coders steal good code snippets from programs that are already written and known to work the way you want.

Now, after everything finally works and you have lots of self-esteem and confidence as a result, you can write programs and see them execute and produce output. You have circumvented some exclusive privileged access. You have succeeded with the hardest task in this book, the rest is all coasting down-hill, giving you the freedom to think without distractions.

Chapter 8

Computer Code for an ASCII Demo

The following example code produces a print-out of the alphabet, numbers, and several symbols. These characters are the interface between humans and computers when using a keyboard or reading plain text on a computer screen. In addition to the keyboard and the screen, these symbols are stored in files and transmitted inside computers talking with and to themselves, in other words, the numbers that represent these symbols are how the computer reads and writes both to itself and to the human world. This is the simplest and oldest set of symbols for computer communication purposes. Today, especially with the Internet, other more extensive sets of symbols are needed.

The symbols here are called ASCII characters, American Standard Code for Information Interchange, where each symbol on the computer keyboard is associated with a number, called the ASCII numeric representation of a symbol. For example, the upper-case letter A is represented by the number 65. All this becomes obvious when examining the output of the following computer code.

Since these are among the first examples of computer code in this book, they are short and simple, intended to meet the needs of the person who sees all this for the first time and with very limited experience with computers, so that nothing can be considered too simple or obvious. Failure is common but never final when working with computers, the cause usually is overlooking, not remembering, or not knowing one simple detail.

With the text editor, write and save the four files with contents shown below. Often but not always, the space-character is very important for the computer to understand what computer instructions you are trying to

CHAPTER 8. COMPUTER CODE FOR AN ASCII DEMO

write. In particular, in Python programs, the space-characters including the tab-character at the beginning of a line indicate how lines are grouped together, and, any carelessness will result in the Python program failing to run. The C language and the Fortran language are more forgiving regarding space-characters at the beginning of a line but are less forgiving in other ways. Handling space-characters may seem to be an idiosyncrasy, but, computer programming is little more than an orderly set of idiosyncrasies.

```c
/* this file name:    C_8_ascii.c
 * dependencies: none
 * to compile:   gcc -o D_8_run_c.out C_8_ascii.c
 * to execute:   ./D_8_run_c.out
 * This example code illustrates:
 *    print ASCII characters and numbers
 */
#include <stdio.h>
main()
{
   int j;
   for(j = 32; j < 130; j++)
   {
           printf(" %d %c\n",j,j);
   }
   return 0;
}
```

```fortran
! this file name:    C_8_ascii.f90
! dependencies: none
! to compile:
!      gfortran -o D_8_run_f90.out C_8_ascii.f90
! to execute:   ./D_8_run_f90.out
! This example code illustrates:
!    print ASCII characters and numbers
   program ASCII
   do 10 j = 32, 129
      write(*,*)j ,CHAR(j)
10    continue
   end
```

CHAPTER 8. COMPUTER CODE FOR AN ASCII DEMO

```
# this file name:   C_8_ascii.py
# dependencies: none
# to execute:  python C_8_ascii.py
# This example code illustrates:
#    print ASCII characters and numbers
for i in range(32,130):
    a = chr(i)
    print(a),;print(" "),;print(i)
```

```
# this file name:   C_8_ascii_doit.py
# dependencies:   C_8_ascii.py
#                 C_8_ascii.c
#                 C_8_ascii.f90
# to execute:  python C_8_ascii_doit.py
# This example code illustrates:
#              using Python to automate tasks
import os
os.system('python C_8_ascii.py > E_8_see_py.txt')
os.system('gcc -o D_8_run_c.out C_8_ascii.c')
os.system('./D_8_run_c.out > E_8_see_c.txt')
os.system('gfortran -o \
          D_8_run_f90.out C_8_ascii.f90')
os.system('./D_8_run_f90.out > E_8_see_f90.txt')
```

The extension to the filename indicates the type of the file. The extensions .c and .py and .f90 indicate the programming languages C and Python and Fortran, respectively.

The last file, C_8_ascii_doit.py, does everything in this demo, that is, it compiles what is needed and executes the three programs in the first three files, above. This may be seen in the lines that start with *os.system*, which is a Python language instruction to the computer system to do a task, which is given inside the parentheses that follow the instruction. The first task given to the computer system is *python C_8_ascii.py > E_8_see_py.txt*, which means, execute the application named python, which executes the program contained in the file named C_8_ascii.py, and creates and fills an output file named E_8_see_py.txt. The instructions to the computer system in

CHAPTER 8. COMPUTER CODE FOR AN ASCII DEMO   45

the next line are *gcc -o D_8_run_c.out C_8_ascii.c*, which means, execute the compiler called gcc, creating an executable file named D_8_run_c.out, and using as input the code file named C_8_ascii.c. The next line, *./D_8_run_c.out > E_8_see_c.txt*, tells the system to execute the file named D_8_run_c.out and put the output into a file named E_8_see_c.txt. The next two lines do the same things as above for the Fortran program located in the file named C_8_ascii.f90.

Continuing with the Python code, the code inside the parenthesis immediately following os.system is called in computer lingo, the "argument". In this case, the argument is the instructions we want the computer system to execute, and, the argument is in a different computer language, called bash. When you are controlling the computer in Terminal mode, you are using the bash language. If you code, you must expect to run into another computer language now and then, so, you learn a minimum and keep a cheat sheet handy.

To execute this demo, in Terminal mode and in the directory containing the above four files, enter with the keyboard: *python C_8_ascii_doit.py* and press the Enter-key. Three files with output are created. These three output files may be examined by opening them with the text editor, actually when in File Manager just click on them and the computer knows you want to open that file, and, the computer sees the name extension, .txt, so opens the file with the text editor. What will be seen in each are the character symbols and the ASCII number that represents each symbol.

This completes all the instructions for setting up hardware and software necessary for coding, that is, writing and executing computer programs, and examining the resulting output. Using a sports analogy, now that we have a soccer field and a soccer ball, and we can successfully kick the ball, it's time to turn our attention to the intricacies of game strategies.

Chapter 9

Other Options

Before we get very far down the road, the devil's advocate comes to ask, why not do water quality assessment using something more familiar than Bayes Theorem, such as an average of all the water quality monitoring observational data. The questioner really is trying to simplify things, make the process easier to understand, and still pass the laugh test. This is important to consider, because water quality assessment is more about attitudes, preconceived notions, financial ramifications, things that are not usually associated with science and the perceived focus of science to understand the true facts of the world and other areas of interest. Water quality assessment is more closely aligned with preponderance of the evidence than any desire to discover and prove a fact scientifically, which is difficult at best and nearly impossible for quality assessments.

Now, we may think to ourselves that the devil's advocate has not been listening closely, and does not know or remember that we are interested in the situation where the observational data may be a number, but also may be anything such as a word, a paragraph, several numbers, words and numbers together. So, suggesting an average as a better approach really won't work for us for every case we're interested in. We, however, recognize the devil's advocate is often a very useful source of ideas we may have overlooked, or something, and, some one else will likely require us to address the same issue, and some one else may be influential enough and powerful enough to cause us great anxiety if we don't respond adequately. So we always respond to the devil's advocate.

The following computer program illustrates two approaches to take a sequence of observations, presumably gathered at different times and all gathered from the same water body of interest. One approach is a ten point rolling average, which pleases the devil's advocate for thinking of

CHAPTER 9. OTHER OPTIONS

it first, the other approach is Bayes Theorem used as an update process. The devil's advocate takes the lead and argues that the average of all the data is a pretty good way to track a variable, especially if the process being monitored is constant and consistent in time. But what if the process monitored is not constant and consistent in time. The devil's advocate continues with recommending a ten point rolling average, which is better than the average of the total data, because the rolling average follows only the most recent observations, and will reasonably quickly settle on the average of the current conditions of the process being monitored. Ten points is suppose to smooth out bumps and dips that would lead a more responsive averager to make a false conclusion, but, not be as unresponsive as the average of the total, avoiding failing to notice a true change in the process. Great. Take on the devil's advocate and find out, in this case, how the behavior of the ten point rolling average compares with a Bayes Theorem approach.

The door opens with the following in-depth reading of computer code, the narratives added to the code here are enclosed with the C convention for multi-line comments, which ignores everything enclosed by /* and */. The narratives give the overall organization and rational, the details of the C programming language are better learned else where. Everything below is code that will compile with out error and execute on the frugal salvage old laptop installed with Linux at no cost, as described above.

To follow computer code, start reading at the first character in the file, and, read one character at a time in order from beginning to end; this is what the computer does. At the same time, read the "words" that appear consecutively, and if well written the human mind finds it is readable and understandable. Start:

```
// this file name:    C_9_step_annotated.c
// dependencies:   none
// to compile:
//      gcc -o C_9_run_c.out C_9_step_annotated.c
// to execute: ./C_9_run_c.out
// This example code illustrates:
//         demonstrate step-function side by side,
//         ten point rolling average, and
//         Bayes update

/* Each computer language has its own way of
 * doing business, here the "#include" needs
```

CHAPTER 9. OTHER OPTIONS 48

```
 * to appear before any other code in order to
 * import code previously developed to do
 * particular tasks. In this case, stdio.h is a
 * file containing C code that does all types of
 * input and output tasks; you can see how
 * complexity results in layers upon layers, and,
 * if you wish you can open the file stdio.h and
 * see the C code it contains. The original
 * intent was to make the programmer's life easier
 * and more productive. Later it was realized
 * that this and similar approaches are a great
 * way to hide details in plain sight, and, many
 * are good examples of obfuscated code which
 * results in exclusive privileged access.
 */

#include <stdio.h>

/* The computer must have some way to know where
 * to start, a program written in C always
 * indicates the starting point with the following
 * line. Later extensions to C expand on this
 * simple line, adding complexity.
 */

main()

/* Code has structure, that is, a way in which it
 * is organized so the computer knows what to do
 * next. The open curly bracket signals that
 * everything that follows the bracket, until a
 * closing curly bracket appears, is all executed
 * together in the order of appearance. Curly
 * brackets can be nested inside each other. This
 * first open curly bracket starts the enclosure
 * of this entire computer program, which ends at
 * the last closing curly bracket.
 */

{

/* All variables must be declared before they are
```

CHAPTER 9. OTHER OPTIONS

```
 * used.
 */

    int j, k, ntot;
    float prior, offset, prob, tot,
/* This current line of code, written in C, does
 * not end until the ; appears, so it may spill
 * over to a new line.
 */
    ave, xprior, xprob, pnum, pden, post;

/* Initialize variables before they are used. */

    prior = 0.5;

    ntot = 0;
    tot = 0.0;

    offset = 0.3;
    prob = 0.5 + offset;


/* Here set up a loop, and the open curly bracket
 * indicates the start of a sequence of code that
 * is executed together. This sequence enclosed in
 * curly brackets is executed over and over, until
 * some criterion is satisfied.
 */

    for( j = 1; j < 1000; j++ )
    {

/* First in this sequence is the task of creating
 * the data which will be processed in a ten point
 * rolling average and by Bayes update. Here a
 * datum is the value of the variable named prob.
 */

// step-function
        k = j % 40;
        if(k == 0)
        {
```

```
        offset = -offset;
        prob = 0.5 + offset;
    }

/* Second in the sequence is the task of
 * accumulating the ten point rolling average,
 * using addition, multiplication, and division
 * of the values of several variables. Here the
 * ten point rolling average is the value of the
 * variable ave, and the data are the values of
 * the variable prob.
 */

// ten point rolling average
    if( j <= 10)
    { /* here if less than ten datum */
        ntot = ntot + 1;
        tot = tot + prob;
        ave = tot / ntot;
    }
    else
    {/* here if enough data for ten points */
        ave = ( (ave * 10.0) + prob ) / 11.0;
    }

// Bayes update
        /* see other chapters for details */
        xprior = 1.0 - prior;
        xprob  = 1.0 - prob;

        pnum = prob * prior;
        pden = pnum + xprob * xprior;

/* post is the variable containing the result,
 * and here is considered suitable for comparison
 * purposes with the ten point rolling average.
 */
        post = pnum / pden;

/* The way Bayes Theorem works, if the value of
 * post becomes zero or one, then the process
 * locks up.
```

CHAPTER 9. OTHER OPTIONS 51

```
      */
        if(post > 0.99)post = 0.99;
        if(post < 0.01)post = 0.01;

        prior = post;
// write
   /* Here write the results so we can see them. */
        printf(" %d %f %f %f\n",j,prob,ave,post);

   /* This curly bracket is the end of the loop.*/
        }

/* This curly bracket is the
 * end of the computer program. */
}
```

```
# this file name:   C_9_step_doit.py
# dependencies:  C_9_step.c
# to execute:   python C_9_step_doit.py
# This example code illustrates:
#                using Python to automate tasks
# NOTE:  we may do the same for
#                C_9_step_annotated.c
#        which should give identical output
import os
os.system('gcc -o D_9_run.out C_9_step.c')
os.system('./D_9_run.out > E_9_see.txt')
```

```
/* this file name:   C_9_step.c
 * dependencies: none
 * to compile:
 *         gcc -o C_9_run.out C_9_step.c
 * to execute:   ./C_9_run.out
 * This example code illustrates:
 *     step-function with side by side
 *          ten point rolling average, and
```

```
 *                 Bayes update
 */
#include <stdio.h>
main()
{
        int j, k, ntot;
        float prior, offset, prob, tot,
        ave, xprior, xprob, pnum, pden, post;
    prior = 0.5;
    ntot = 0;
    tot = 0.0;
    offset = 0.3;
    prob = 0.5 + offset;
    for( j = 1; j < 1000; j++ )
    {
// step-function
        k = j % 40;
        if(k == 0)
        {
            offset = -offset;
            prob = 0.5 + offset;
        }
// ten point rolling average
        if(j <= 10)
        {
            ntot = ntot + 1;
            tot = tot + prob;
            ave = tot / ntot;
        }
        else
        {
            ave = ( (ave * 10.0) + prob ) / 11.0;
        }
// Bayes update
       xprior = 1.0 - prior;
       xprob  = 1.0 - prob;
       pnum = prob * prior;
       pden = pnum + xprob * xprior;
       post = pnum / pden;
       if(post > 0.99)post = 0.99;
       if(post < 0.01)post = 0.01;
       prior = post;
```

```
// write
    printf(" %d %f %f %f\n",j,prob,ave,post);
    }
}
```

As can be seen when examining the output, the ten-point rolling average is slower to indicate that a step in the data has occurred. By providing this demonstration, where every detail can be examined, and every detail may be tweaked by changing the computer program, the devil's advocate has the opportunity to gain more or less understanding of what is really happening with a ten point rolling average, and with a Bayes update process. This understanding allows the human mind to form an estimate of the truth of various aspects of the understanding, and arrive at a point, eventually, of being able to make a decision for some question when presented data, observations, methods of some process. No mention of "proof" here, computer programming allows the human mind to tally, organize, manipulate, anything representable by numbers, and observe the results of processes that completely overwhelm the human mind thinking without outside aid. Proof in mathematics is narrowly defined and contrasts to computer programming where any way to throw numbers and symbols around is allowable as long as it compiles and executes. Proof in the physical, biological, economics, social and psych sciences, and all the other sciences, all have their norms for what could be called proof, but, in reality is a state of understanding that has considerable consistency and agreement among measurable and observable quantities and various theories that attempt to describe these observed behaviors. A simplistic example, has anyone mathematically proved gravity? Not yet. But that does not matter because understanding gravity is important for getting around in every day life. What about "truth"? Gravity is true, because we understand gravity's existence and we don't go around casually jumping off tall buildings. "Truth" in our minds is important because we need it to survive, both as individuals and as a species. If we can't wrap our minds around gravity, we can explore how gravity behaves, either classically with pencil, paper, and our minds, or, we can explore with keyboard, screen, and our minds. Computer technology allows our minds to explore way beyond what we can wrap our minds around even with the powerful aid of pencil and paper.

The above quite short and quite simple computer code that compares the behavior side by side of a ten point rolling average and a Bayes update process illustrates what this book is intended to do, that is, assist people to get, frugally, simply, and quickly, the where-with-all to use computer programming to assist in thinking about issues and situations that are not possible for the average person to grasp in their minds alone.

Chapter 10

Reasonable Expectations

Bayes Theorem, used in an update process to get the most recent estimate about the status of water quality for a particular segment and applicable standard, is hyped here as being the fastest and best possible, any other approach may be as fast and as good, but never faster or better. This hype can easily turn into unrealistic expectations. The following computer code is intended to offer sobering evidence that Bayes Theorem can not do magic nor the impossible, and the same is true for any analysis of any data obtained from observations and measurements of events influenced by randomness and using techniques based on statistics and probability.

But when using Bayes Theorem, more sobering is that a seemingly small difference in wording can cause a major difference in performance, all else being equal. Later, an entire chapter will be devoted to the place in the code where this seemingly small change is made, and it will be given a name, in Chapter 17 it is called a translator.

The following portrays two players of a new dice game. One player is dishonest, that is, manages with slight of hand to swap an honest pair of dice with a pair that is loaded to favor the dishonest player. The question is, how long will it take the other presumed honest player to figure out that the game is somehow modified so that in the long run, one player will win and the other will lose. The answer is much longer than would be wished for, and with a different wording, much better but not outstanding.

The following is annotated source code for simulating a dishonest dice game, that will compile without errors and execute as intended.

```
// this file name:    C_10_dice.c
```

CHAPTER 10. REASONABLE EXPECTATIONS

```c
// dependencies:   none
// to compile:   gcc -o C_10_run.out C_10_dice.c
// to execute: ./C_10_run.out
// This example code illustrates:
//      A dice game, showing the abilities
//      and limitations of statistics, and,
//      the necessity of computer literacy.
#include <stdio.h>       // So we can print
#include <stdlib.h>      // So we can use rand()
                         //         and srand().
main()    // Every C-program starts with this.
{
/*
 * Two people, Shark and Stooge, play a new
 * street game of dice. Stooge does not know
 * that Shark uses dishonest dice. How long
 * will it take for Stooge to realize, and
 * conclude the game is dishonest and
 * in Shark's favor?
 */
    // Declare variables used here.
    int i, j, itot, jtot, ida, idb, ix;
    float shark, stooge, pillow, diff, uneven;
    float prior, prob, xprior, xprob, pnum, pden,
                                            post;

/* Do the following twice,
 * make one small difference,
 * first, Stooge accepts the evidence as obvious,
 *      which is, if losing consistently then
 *      the game is dishonest,
 * second, Stooge uses the evidence slightly
 *      differently, by including consideration
 *      of available knowledge.
 */
    j = 1;
    while ( j <= 2 )
    {

    srand( 12345 ); /* Initialize the random
                     * number generator. To run
                     * a different scenario, edit
```

CHAPTER 10. REASONABLE EXPECTATIONS 57

```
                    * this computer code changing
                    * 12345 to some other
                    * integer.
                    */


         shark = 0.0;    // Initialize the winnings
                         // for Shark.
         stooge = 0.0;   // Same for Stooge.
         pillow = 0.0;   // What Stooge calculates
                         // should be in Stooge's pillow
                         // (used in second pass).

         prior = 0.5;    /* Initialize the Bayes prior,
                          * 0.5 means there is no
                          * evidence for believing
                          * either conclusion, honest
                          * or not.
                          */

// Start playing, for a long time.
         for( i = 1; i < 10000; i++ )
           {
//  Roll two dice, add them together, get result,
//  range 2 through 12.
                 ida = rand() % 6 + 1;  // An honest
                                        // dice, value
                                        // is 1 through 6,
                                        // each equally possible.
                 idb = rand() % 6 + 1;  // Another
                                        // honest dice.
                itot = ida + idb; // Result of rolling
                                  // two honest dice together,
                                  // a number between 2 and 12
                                  // inclusively.

/* But the dice in this illustration are suppose
 * to be dishonest, so redo by rolling the two
 * dice again. The above is just an example of
 * an honest roll.
 */
```

CHAPTER 10. REASONABLE EXPECTATIONS   58

```
            ida = rand() % 6 + 1; // start with
                                  // honest roll.

            jtot = rand() % 6 + 1; // Stooge will
                    // use this in the second pass.

            ix = rand() % 1000;
            if( ix >= 500 )
            {   /* Make this honest roll dishonest
                 * as follows: half the time the
                 * result is one less than a result
                 * of an honest dice.  In other
                 * words, half the time the
                 * dishonest dice rolls only a 1
                 * through 5, never 6, and a 1
                 * twice as likely than the others.
                 * (Definitely not a dice
                 * fabricated in the real world.)
                 */
                ida = ida - 1;
                if( ida == 0 ) ida = 1;
                  // Now this is a dishonest dice.
            }

            idb = rand() % 6 + 1; // Honest,
                                  // leave it.

            itot = ida + idb; /* The result of the
                 * roll of the two dice,
                 * one honest, the other dishonest;
                 * a number between 2 and 12
                 * inclusively.
                 */

            jtot = jtot + rand() % 6 + 1;
               // Stooge will use this in
               // the second pass.

/*
 * How this game moves the money:
 * 1.  If the two dice roll a 7, take another
 *     turn, no money exchanged.
```

CHAPTER 10. REASONABLE EXPECTATIONS

```
*  2.   If the roll is greater than 7, Stooge
*       wins and gets money from Shark, the amount
*       of money per play is agreed beforehand, say
*       one dollar.
*  3.   If the roll is less than 7, Shark wins.
*  4.   If Stooge wants to flip the rules, then
*       Shark agrees and with slight-of-hand
*       replaces one dishonest dice for another
*       that is in Shark's favor.
*       (This complication is not simulated here.)
*  5.   But more, Shark takes advantage of Stooge's
*       human tendency of greed, and offers uneven
*       payoff, so, Stooge presumes honest dice and
*       concludes Stooge will win in the long run
*       because of the uneven payoff, and because
*       Shark is just a nice person and doesn't
*       care about losing or winning.
*/
            uneven = 1.15; /* Shark's offer of
                * uneven payoff, not part of the game
                * but added by Shark because Shark
                * believes that Stooge is not too
                * sharp and will play forever with
                * the incentive. */

// The dice are rolled, move the money,
// one dollar per roll of the dice.
            if( itot > 7 )
            {
                stooge = stooge + uneven;
                shark  = shark  - uneven;
                // Money moves unevenly
                // in Stooge's favor.
            }
            if( itot < 7 )
            {
                stooge = stooge - (1.0 / uneven);
                shark  = shark  + (1.0 / uneven);
                // Again, uneven payoff.
            }
// No money moves if itot equals 7.
```

CHAPTER 10. REASONABLE EXPECTATIONS   60

```
/* Now look at two ways for Stooge to evaluate
 * the evidence. For each way the result answers
 * the same question:
 * What is the likelihood that Shark is playing
 * an honest game. */
            if( j == 1 )  // First pass.
            {
/* Stooge uses common sense and sees the obvious,
 * and Stooge looks only at Stooge's winnings and
 * believes that if the total winnings
 * are positive in the long run then the
 * game is honest and the positive winnings
 * are the result of the uneven payoff. */
                if(shark < stooge)
                    prob = 0.6; // Stooge is
                                // always skeptical
                else
                    prob = 0.4; // but skeptically
                                // believes what
                                // appears obvious.
            }

            if( j == 2 )  // This is the
                          // second pass.
            {
/*  Here Stooge is less naive about what appears
 *  obvious; Stooge calculates how the money
 *  should move if the game is honest, and,
 *  including uneven payoff. */
                if ( jtot > 7 )
                    pillow = pillow + uneven;
                if( jtot < 7 )
                    pillow = pillow -
                                (1.0 / uneven);
/* Stooge calculates the expected winnings as if
 * the game is honest and the payoff is uneven
 * and believes that if the actual total winnings
 * are the same as the calculated winnings in
 * the long run then the game is honest. */
                diff = pillow - stooge;
                if(diff <= 0.0)
                    prob = 0.6; // Stooge always
```

CHAPTER 10. REASONABLE EXPECTATIONS 61

```
                        // skeptical,
        else
            prob = 0.4; // but skeptically
                        // believes what
                        // appears obvious.
    }

/* prob is Stooge's guess and gut feeling about
 * the game honesty, based on current evidence
 * that Stooge can see, and here two different
 * ways of seeing and believing the evidence are
 * explored. The way that prob is calculated will
 * establish the way the question is worded, do
 * we want the likelihood of honest, or the
 * likelihood of dishonest.
 */


// Bayes update, exactly the same for both passes:

            xprior = 1.0 - prior;
            xprob  = 1.0 - prob;

            pnum = prob * prior;
            pden = pnum + xprob * xprior;

            post = pnum / pden; /* This is the
             * likelihood, expressed as probability,
             * that Shark and the dice are both
             * honest. */

            if(post > 0.99)post = 0.99;
            if(post < 0.01)post = 0.01; /* These
                             * hard limits
                             * are necessary to
                             * prevent problems of
                             * latching at zero
                             * or one. */

            prior = post; // Update for the
                          // next round.
```

```
            // Write the progress, cumulative.

                        printf(" %d %d %d %f %f %f %f\n",
                                    j,itot,jtot,pillow,
                                    stooge,shark,post);

/*
 * Using Bayesian approach to assist decision
 * making, Stooge looks at the value of post as
 * follows:
 *     Using only the evidence and understanding
 *     of the game, and nothing more,
 *     If post equals 0.01 then Stooge can
 *         assuredly conclude the game is
 *         dishonest in favor of Shark,
 *     If post equals 0.99 then Stooge can
 *         assuredly conclude the game is not
 *         dishonest in favor of Shark,
 *     If post equal 0.5 then Stooge has no
 *         evidence either way about the game
 *         dishonesty.
 * Stooge should stop playing dice with Shark
 * when post reaches 0.01, or sooner if Stooge
 * prefers.  If post reaches 0.99, or sooner,
 * Stooge may continue playing with Shark without
 *  worrying too much about losing money.
 */

            }   // End of a pass.

            j = j + 1;   // Increment the pass
                         // over two ways to examine evidence.

        }   // End of all passes.

    return;
}   // End of this program.
```

```
# this file name:   C_10_dice_doit.py
```

CHAPTER 10. REASONABLE EXPECTATIONS

```
# dependencies:    C_10_dice.c
# to execute:    python C_10_dice_doit.py
# This example code illustrates:
#            using Python to automate tasks
import os
os.system('gcc -o D_10_run.out C_10_dice.c')
os.system('./D_10_run.out > E_10_see.txt')
```

In general, the human mind is astute at recognizing patterns and correlations in data. However, the astuteness is not necessarily reliable. The above code illustration takes two passes through the same data. In the first pass, the mind is looking for something obvious, which is, if one player consistently wins over some reasonable length of time then the astute mind latches on to the conclusion that the game at least is not dishonest for that player.

This has significant implications for using Bayes Theorem, if the process is set up casually or carelessly or with insufficient attention to details, then, the Bayes Theorem approach will likely not perform as desired, and, may be discarded as inferior, when the fault is actually not Bayes Theorem but how Bayes Theorem is used. This of course may be generalized to any approach to any issue, the human mind evolved to respond very well to the environment and social bondings necessary for survival of the species, but not as good for more abstract or more complex situations. Consider Einstein's general relativity, how many people feel they understand it, compared to understanding what happens when a pitched base ball is hit by a base ball bat.

After examining the output of the first pass through the above computer code, the reader may decide that neither the human mind nor Bayes update give the results wished for, which is a quick simple definitive recognition that the other player is dishonest. This should temper any expectation that truth and understanding will be easy with real world data that are confounded by randomness, unknown factors, and noise.

On the other hand, the second pass shows how powerful Bayes Theorem can be, when used in a manner that utilizes the capabilities inherent in Bayes Theorem which in the example here means use all the information available. This in general includes attention to details, careful wording of the question, and understanding as best as possible the true situation being investigated.

When dealing with assessment and quality, which are abstract and depend on human emotion and beliefs more than depend on skills and knowledge, successful use of Bayes Theorem will involve a continuous process of implementation, evaluation, and revision, as the situation being addressed is better understood with each improvement cycle. This of course is thinking augmented with assistance from computers.

Chapter 11

Choose the Programming Language

Thus far, three programming languages have been used to write complete examples of computer coding. In this chapter, the goal is that one will be selected as the primary language used for writing the code that does the heaviest work. The following is a test program that gives the same work to three different computer languages plus a capability provided by the operating system, and we will observe which of the four executes the work in the shortest time. Presumably we select the fastest, in spite of and cognizant that later evidence may cause us to change our mind. Think of this as an exercise in how to proceed when we find ourselves in a similar situation in the real world.

Selection of the work, also called a benchmark, is not arbitrary. The approach to assessment here will be a computer program that spends most of the execution time on one sub-task, and comparatively trivial time on everything else. There are two limitations that will limit the usefulness of the computer program, one is the time it takes to convert data into results, and the second is the size of the data which must be not larger than the computer can handle.

The data here will involve water bodies, and beneficial uses for each water body, and standards for each beneficial use for each water body. Depending on how these three are defined, the data could easily contain millions of entries, for the contiguous 48 states, including rivers and streams, coast lines, lakes, impoundments, diversions, estuaries, basins, wetlands, tidelands, et cetera.

The critical work here is sorting a list of data, which as an aside, is a

CHAPTER 11. CHOOSE THE PROGRAMMING LANGUAGE 66

way to bring order to chaos. The particular approach used here is called heap sort, details of which appear in many books and the Internet, and a System application named appropriately, sort. The particular heap sort algorithm used here is taken from Numerical Recipes, a book that is viewable on the Internet.

```
# this file name:   C_11_compare_doit.py
# dependencies:    C_11_compare_running.f90
# buried dependencies: C_11_data.py
#                     C_11_rrsort.py
#                     C_11_qqsort.c
#                     C_11_ppsort.f90
#                     C_11_ttsort.py
# to execute:   python C_11_compare_doit.py
# This example code illustrates:
#          using Python to automate tasks
import os
os.system('gfortran -o D_11_run.out \
              C_11_compare_running.f90 ')
os.system('./D_11_run.out')
```

```
! this file name:   C_11_compare_running.f90
! dependencies:    C_11_data.py
!                  C_11_rrsort.py
!                  C_11_qqsort.c
!                  C_11_ppsort.f90
!                  C_11_ttsort.py
! to compile:
!      gfortran -o D_11_run_all.out
!                 C_11_compare_running.f90
! to execute:   ./D_11_run_all.out
! This example code illustrates:
!       find running times for an algorithm using
!       several programming languages
         program C_11_compare_running

         write(*,*)'start'
         call system_clock(j1)
```

CHAPTER 11. CHOOSE THE PROGRAMMING LANGUAGE 67

```
      call system('python C_11_data.py')
      write(*,*)'data complete, start Python sort'
      call system_clock(j2)

      call system('python C_11_rrsort.py')
      write(*,*)   &
     'Python sort complete, start system sort'
      call system_clock(j3)


      call system    &
     ('sort E_11_CSrandom.txt > E_11_CSsortedS.txt')
      write(*,*)'system sort complete'
      call system_clock(j4)

      call system    &
     ('gcc -o D_11_CSqq.out C_11_qqsort.c')
      write(*,*)'                  start C sort'
      call system_clock(j5)

      call system('./D_11_CSqq.out')
      write(*,*)'C sort complete'
      call system_clock(j6)

      call system    &
     ('gfortran -o D_11_CSpp.out C_11_ppsort.f90')
      write(*,*)'            start Fortran sort'
      call system_clock(j7)

      call system('./D_11_CSpp.out')
      write(*,*)'Fortran sort complete'
      call system_clock(j8)

      call system('python C_11_ttsort.py')
      write(*,*)   &
     'second be fair to Python sort complete'
      call system_clock(j9)

      f = 0.001
      open(11,file='E_11_seeresults.txt')
      write(11,*)'Python time   = ',REAL(j3-j2) * f
```

```fortran
        write(11,*)'System       = ',REAL(j4-j3) * f
        write(11,*)'C            = ',REAL(j6-j5) * f
        write(11,*)'Fortran      = ',REAL(j8-j7) * f
        write(11,*)'Python again = ',REAL(j9-j8) * f
        close(11)

        end
```

```python
# this file name:   C_11_data.py
# dependencies: none
# to execute:   python C_11_data.py
# This example code illustrates:
#    generate simulation data, random
import random
import os
file_a = open("E_11_CSrandom.txt", mode = "wb")
# NOTE: indenting lines determines the structure
#       of the code, the indent white space is
#       a detail of Python that is unforgiving
#       if the coder is not careful
for i in range(1,500000):
        j = random.random()
        k = random.random()
        file_a.write('{:014.12f}'.format(k) + \
             " " + '{:06d}'.format(i) + \
             " " + '{:014.12f}'.format(j) \
             + " AAA\n")
file_a.close()
```

```fortran
! this file name:   C_11_ppsort.f90
! dependencies: none
! to compile:   gfortran -o D_11_CSpp.out
!                         C_11_ppsort.f90
! to execute:   ./D_11_CSpp.out
! This example code illustrates:
!    heap sort from Numerical Recipes pg. 329,
!    altered for character strings
```

CHAPTER 11. CHOOSE THE PROGRAMMING LANGUAGE 69

```fortran
      program ppsort
      character*256 fin,fou
      character cuf
      fin = 'E_11_CSrandom.txt'
      fou = 'E_11_CSsortedF.txt'
      INQUIRE(FILE=fin,SIZE=lengf)
      nin = 12
      open(nin,file=fin)
      cuf = char(32)
      lenrec = 0
      do while (ichar(cuf) .NE. 10)
         call fgetc(nin,cuf,istat)
         if(istat.NE.0)go to 999
         lenrec = lenrec + 1
      end do
      close(nin)
      numrec = lengf / lenrec
      call sszort(lenrec,numrec,fin,fou)
999   continue
      end
```
!————————————————————————————————
```fortran
      subroutine sszort(lenrec,numrec,fin,fou)
      character(lenrec) rra
      character*(*) fin, fou
      character(lenrec), allocatable :: ra(:)
      character cuf
      allocate(ra(numrec))
      nin = 12
      open(nin,file=fin)
      narr = 0
110   continue
      read(nin,'(a)',end=120)rra
      narr = narr + 1
      ra(narr) = rra
      go to 110
120   continue
      close(nin)
      n = narr
      ll = n / 2 + 1
      ir = n
      do while(ir .GT. 0)
         if(ll.gt.1)then
```

```
                ll = ll - 1
                rra = ra(ll)
            else
                rra = ra(ir)
                ra(ir) = ra(1)
                ir = ir - 1
                if(ir.eq.1)then
                    ra(1) = rra
                    go to 30
                end if
            end if
            i = ll
            j = ll + ll
            do while(j .LE. ir)
                if(j.lt.ir)then
                    if(ra(j).lt.ra(j+1))then
                        j = j + 1
                    end if
                end if
                if(rra.lt.ra(j))then
                    ra(i) = ra(j)
                    i = j
                    j = j + j
                else
                    j = ir + 1
                end if
            end do
            ra(i) = rra
        end do
30      continue
        nin = 12
        open(nin, file=fou)
        do 230 i = 1, narr
            rra = ra(i)
            le = LEN_TRIM(rra(1:lenrec))
            write(nin,'(a)')rra(1:le)
230     continue
        close(nin)
        return
        end
```

CHAPTER 11. CHOOSE THE PROGRAMMING LANGUAGE

```c
// this file name:    C_11_qqsort.c
// dependencies:  none
// to compile:    gcc -o D_11_CSqq.out C_11_qqsort.c
// to execute:    ./D_11_CSqq.out
// This example code illustrates:
//    heap sort from Numerical Recipes pg. 329,
//    altered for character strings
#include <stdio.h>
#include <stdlib.h>
#include <sys/stat.h>
int main()
{
   int kount;
   FILE * fp;
   FILE * fq;
   struct stat st;
   long int size;
   int j, lenrec, manyrecs;
   char *b;
   char *t;
   int n,ll,i,ir,loop;    int ks, ks1, ks2;
    int k,m,ic;
    stat("E_11_CSrandom.txt", &st);
    size = st.st_size;
    b = (char*) malloc( sizeof(char) * size );
    fp = fopen( "E_11_CSrandom.txt" , "r" );
    if( fp != NULL )
    {
       fread(b,size,1,fp);
       fclose( fp );
       lenrec = 0;
       for(j=0;j<size;j++)
       {
          if( 10 == (int)(char)b[j])
             {
                lenrec = j + 1;
                break;
             }
       }
          t = (char*) malloc( sizeof(char)
                                    * lenrec);
          manyrecs = size / lenrec;
```

```
n = manyrecs;
ll = n / 2 + 1;
ir = n;
loop = 0;
while ( ir > 0 )
{
   loop = loop + 1;
   if ( ll > 1 )
   {
      ll = ll - 1;
      ks = ll * lenrec - lenrec;
      for ( k = 0; k < lenrec; k++ )
      {
         t[k] = b[k+ks];
      }
   }
   else
   {
      ks = ir * lenrec - lenrec;
      for ( k = 0; k < lenrec; k++ )
      {
         t[k] = b[k+ks];
      }
      ks = ir * lenrec - lenrec;
      for ( k = 0; k < lenrec; k++ )
      {
         b [k+ks]= b[k];
      }
      ir = ir - 1;
   }
   if ( 1 == ir )
   {
      for ( k = 0; k < lenrec; k++ )
      {
         b[k] = t[k];
      }
      break;
   }
   i = ll;
   j = ll + ll;
   while ( j <= ir )
   {
```

```
            if ( j < ir )
            {
                ks1 = j * lenrec - lenrec;
                ks2 = (j+1) * lenrec - lenrec;
                ic = 0;
                for ( k = 0; k < lenrec; k++ )
                {
                    if( b[k+ks1] < b[k+ks2] )
                                      ic = -1;
                    if( b[k+ks1] > b[k+ks2] )
                                      ic = +1;
                    if( 0 != ic) break;
                }
                if ( -1 == ic )
                {
                    j = j + 1;
                }
            }
            ks1 = j * lenrec - lenrec;
            ic = 0;
            for ( k = 0; k < lenrec; k++ )
            {
                if( t[k] < b[k+ks1] ) ic = -1;
                if( t[k] > b[k+ks1] ) ic = +1;
                if( 0 != ic) break;
            }
            if ( -1 == ic )
            {
                ks1 = i * lenrec - lenrec;
                ks2 = j * lenrec - lenrec;
                for ( k = 0; k < lenrec; k++ )
                {
                    b[k+ks1] = b[k+ks2];
                }
                i = j;
                j = j + j;
            }
            else
            {
                j = ir + 1;
            }
        }
```

```
                    ks = i * lenrec - lenrec;
                    for ( k = 0; k < lenrec; k++ )
                    {
                        b[k + ks] = t[k];
                    }
            }
            free(t);
        }
        fq = fopen( "E_11_CSsortedC.txt" , "w" );
        if( fq != NULL )
            {
                fwrite(b,size,1,fq);
                fclose( fq );
            }
        free(b);
}
```

```
# this file name:  C_11_rrsort.py
# dependencies:  none
# to execute:  python C_11_rrsort.py
# This example code illustrates:
#     heap sort from Numerical Recipes pg. 329,
#     altered for character strings
import os
statinfo = os.stat('E_11_CSrandom.txt')
leng = statinfo.st_size
file_b = open("E_11_CSrandom.txt", mode = "rb")
ra = bytearray(file_b.read())
file_b.close
zloc = ra.find(chr(10))
lenrec = zloc + 1
manyrecs = leng / lenrec
n = manyrecs
l = n / 2 + 1
ir = n
while ir > 0:
    if l > 1:
        l = l - 1
        now = l
        ks = (zloc + 1) * (now - 1)
```

CHAPTER 11. CHOOSE THE PROGRAMMING LANGUAGE   75

```
            ke = ks + (zloc + 1)
            rra = ra[ks:ke]
        else:
            now1 = 1
            ks1 = (zloc + 1) * (now1 - 1)
            ke1 = ks1 + (zloc + 1)
            nowir = ir
            ksir = (zloc + 1) * (nowir - 1)
            keir = ksir + (zloc + 1)
            rra = ra[ksir:keir]
            ra[ksir:keir] = ra[ks1:ke1]
            ir = ir - 1
        if ir == 1:
            now1 = 1
            ks1 = (zloc + 1) * (now1 - 1)
            ke1 = ks1 + (zloc + 1)
            ra[ks1:ke1] = rra
            break
        i = 1
        j = 1 + 1
        while j <= ir:
            if j < ir:
                nowj = j
                ksj = (zloc + 1) * (nowj- 1)
                kej = ksj + (zloc + 1)
                nowjp1 = j + 1
                ksjp1 = (zloc + 1) * (nowjp1 - 1)
                kejp1 = ksjp1 + (zloc + 1)
                if ra[ksj:kej] < ra[ksjp1:kejp1]:
                    j = j + 1
            nowj = j
            ksj = (zloc + 1) * (nowj- 1)
            kej = ksj + (zloc + 1)
            if rra < ra[ksj:kej]:
                nowi = i
                ksi = (zloc + 1) * (nowi - 1)
                kei = ksi + (zloc + 1)
                nowj = j
                ksj = (zloc + 1) * (nowj- 1)
                kej = ksj + (zloc + 1)
                ra[ksi:kei] = ra[ksj:kej]
                i = j
```

```
            j = j + j
        else:
            j = ir + 1
    nowi = i
    ksi = (zloc + 1) * (nowi - 1)
    kei = ksi + (zloc + 1)
    ra[ksi:kei] = rra
file_b = open("E_11_CSsortedP.txt", mode = "wb")
file_b.write(ra)
file_b.close
```

```
# this file name:   C_11_ttsort.py
# dependencies:  none
# to execute:   python C_11_ttsort.py
# This example code illustrates:
#     sort, using advanced capabilities of Python
with open("E_11_CSrandom.txt", 'r') as file_a:
    ra = file_a.readlines()
rb = sorted(ra)
file_b = open("E_11_CSsortedPadv.txt", \
                                    mode = "wb")
file_b.writelines(rb)
file_b.close
```

Results located in the file named E_11_seeresults.txt are:
Python time = 62.0430031
System = 4.58500004
C = 3.33300018
Fortran = 1.63300014
be fair Python = 1.22200000

Examining the output, Python appears to be the slowest, the System call next slower, C next, and Fortran is faster. But to avoid the appearance of a rigged test, Python is run again, and this time it is the fastest. When these same computer codes are run on different laptop models and different versions of Linux, the details of the results will differ, but the trend is basically the same, which is, Python is consistently the slowest, and Python is consistently the fastest.

CHAPTER 11. CHOOSE THE PROGRAMMING LANGUAGE 77

Slow Python is easily explained, Python is an interpreter while C and Fortran compile before executing, and, interpreters simply by the way they work are slower. Explaining fast Python, the Python developers realized that number crunching is important and using Python exclusively in interpreter mode for everything is unsatisfactorily slow. So the developers built into Python a fast way to do a specific but important and widely used task, that is, sort. And, it is fast. If we used a different algorithm as a benchmark, and Python does not have a built in capability for that algorithm, then Python will be slow. It may be noticed that no effort was made to test the built-in capabilities to perform sort in C or Fortran, where these built-in capabilities are called libraries.

The System call to sort may be expected to be the fastest, since the people who maintain these System applications see to it that the fastest and best algorithm is used, which here is likely a sort algorithm called timsort, which is faster than heap sort except for completely random data where timsort should run as fast as heap sort. In this test, the System sort is slower than expected. Perhaps the System maintainers and developers are slow in updating the System, perhaps because not enough people want this particular update, perhaps because Crunchbang is no longer supported so uses old slow stuff. So, expect different results for different operating systems and different versions of the same operating system.

The difference between C and Fortran is more difficult to explain, especially when here, gfortran is a front-end pre-compiler that converts Fortran code into C code, which then is compiled with the C compiler. The lesson for us is that nothing is obvious until it is demonstrated. This is part of what understanding and truth are all about. But, in all fairness, more should be said. No effort was expended to make code that runs as fast as possible, so, likely the run time differences between C and Fortran are caused by sub-optimal coding which may be obvious to an experienced programmer, and, if optimized each should be equally fast.

Based on which test ran the fastest, Python wins. However, there are two test entries that do not actually test the fundamental programming language, which are, the System sort, and Python using the built in capability to number crunch sort specifically. Sort is simply our test algorithm, and to conduct our test of programming languages, each test should execute the identical algorithm, which is not the same as executing the same task. So, heap sort is the selected algorithm, and no effort was expended to determine if the System sort or the internal Python sort capability use the heap sort algorithm. And, the result then is that the Fortran code for our purposes, is the fastest, in the test here.

One of our objectives is to write code where the coder is in complete control of all the coding and algorithm details. We started this chapter by saying we will choose a programming language, because the question of which language to use comes up often. We ended by not selecting one language to use, just observing one likely flawed benchmark showed Fortran is fastest for our narrow intended purpose, but when thinking about it, not necessarily generally the best. This chapter is then, an illustration of approaching computer technology. First impressions and preconceived notions often do not survive a close examination based on demonstrative and reproducible evidence. And, any tests of computer language capabilities involve many subtleties, making definitive testing and choosing rather difficult.

Chapter 12

How Well Does Heap Sort Work?

```
# this file name:  C_12_heap_doit.py
# dependencies:   C_12_running.f90
# buried dependencies:
#                 C_12_data.f90
#                 C_12_heapsort.f90
# to execute:   python C_12_heap_doit.py
# This example code illustrates:
#            using Python to automate tasks
import os
os.system \
  ('gfortran -o D_12_run.out C_12_running.f90 ')
os.system('./D_12_run.out')
```

```
! this file name:  C_12_running.f90
! dependencies: C_12_data.f90
!               C_12_heapsort.f90
! to compile:   gfortran -o D_12_run.out
!                  C_12_running.f90
! to execute:   ./D_12_run.out
! This example code illustrates:
```

CHAPTER 12. HOW WELL DOES HEAP SORT WORK?

```fortran
!           automating and timing tasks
          program C_12_running

          write(*,*)'start'
          call system_clock(j1)
          call system    &
          ('gfortran -o D_12_data.out  C_12_data.f90')
          call system    &
       ('gfortran -o D_12_hp.t  C_12_heapsort.f90')
          call system_clock(j2)

          write(*,*)' generate data'
          call system    &
          ('./D_12_data.out   E_12_testdata.txt')
          write(*,*)'        finished'

          write(*,*)' sort data'
          call system_clock(j3)
          call system    &
      ('./D_12_hp.t E_12_testdata.txt E_12_sorted.txt')
          call system_clock(j4)

          open(12,file='E_12_seeresults.txt')
          f = 0.001
          write(12,*)'data generation time= ',  &
                              REAL(j3-j2) * f
          write(12,*)'sort time            = ',  &
                              REAL(j4-j3) * f
          close(12)

          end
```

```fortran
! this file name:   C_12_data.f90
! dependencies: none
! to compile:  gfortran -o D_12_data.out
!                      C_12_data.f90
! to execute:  ./D_12_data.out
! This example code illustrates:
!         generate simulation data
          program C_12_data
```

CHAPTER 12. HOW WELL DOES HEAP SORT WORK?

```fortran
! for the output file-name from the argument
      character*60 arg
      character*256 fou
! for the data
      integer(1), allocatable :: narr(:)
! get the filename from the call argument
      call get_command_argument(1,arg)
      fou = TRIM(arg)
      numlines = 1000000
      lengline = 500
      lengfile = lengline * numlines
      allocate(narr(lengfile))
! initialize by filling with zeros
      narr = 0
! initialize pseudorandom number generator
      iseed = 23579
      x = RAND(iseed)
! fill narr() with ASCII readable data, upper-case
! letters A through Z, put a LF = ASCII 10 at the
! last byte in each record
      do 30 j = 1, numlines
         nstart = lengline * (j - 1) + 1
         nstop  = nstart + lengline - 2
         do 20 k = nstart, nstop
            narr(k) = 65 + INT(RAND() * 26.0)
 20      continue
         narr(nstop + 1) = 10
 30   continue
! save narr() to file
      inquire(IOLENGTH=lengnarr) narr
      open(UNIT=7,FILE=fou,ACCESS='DIRECT',    &
               RECL=lengnarr,ACTION='READWRITE')
      write(7,REC=1) narr
      close(7)
      deallocate(narr)
      end
```

! this file name: C_12_heapsort.f90
! dependencies: none
! to compile: gfortran -o C_12_heap.out

CHAPTER 12. HOW WELL DOES HEAP SORT WORK?

```fortran
!                       C_12_heapsort.f90
! to execute:   ./C_12_heap.out
! This example code illustrates:
!     heap sort from Numerical Recipes pg. 329,
!     altered for character strings
      program C_12_heapsort
!  for the input file-name from the argument
      character*60 arg1, arg2
      character*256 fin, fou
      character cuf
! get the filenames from the call arguments
      call get_command_argument(1,arg1)
      fin = TRIM(arg1)
      call get_command_argument(2,arg2)
      fou = TRIM(arg2)
      write(*,*)'input file name = ',fin(1:40)
      write(*,*)'output file name = ',fou(1:40)
! get size of the file in bytes
      INQUIRE(FILE=fin,SIZE=lenfile)
      write(*,*)'file size=',lenfile
! find the length of a record in the input file,
! presume all records have same length
      nin = 12
      open(nin,file=fin)
      cuf = char(32)
      lenrec = 0
      do while (ichar(cuf) .NE. 10)
         call fgetc(nin,cuf,istat)
         if(istat.NE.0)then
            write(*,*)'ERROR, bad FGETC()'
            go to 999
         end if
         lenrec = lenrec + 1
         if(lenrec .GT. lenfile)then
            write(*,*)'ERROR, no LineFeed in file '
            go to 999
         end if
      end do
      close(nin)
      write(*,*)'record length=',lenrec
      numrec = lenfile / lenrec
      write(*,*)'number records=',numrec
```

CHAPTER 12. HOW WELL DOES HEAP SORT WORK?

```
              call sszort(lenrec,numrec,fin,fou)
999       continue
          end
!  ─────────────────────────────────────────────
          subroutine sszort(lenrec,numrec,fin,fou)
          character(lenrec) rra
          character*(*) fin, fou
          character(lenrec), allocatable :: ra(:)
          character cuf
          allocate(ra(numrec))
          nin = 12
          open(nin,file=fin)
          narr = 0
110       continue
          read(nin,'(a)',end=120)rra
          narr = narr + 1
          ra(narr) = rra
          go to 110
120       continue
          close(nin)
          n = narr
          ll = n / 2 + 1
          ir = n
          do while(ir .GT. 0)
              if(ll.gt.1)then
                  ll = ll - 1
                  rra = ra(ll)
              else
                  rra = ra(ir)
                  ra(ir) = ra(1)
                  ir = ir - 1
                  if(ir.eq.1)then
                      ra(1) = rra
                      go to 30
                  end if
              end if
              i = ll
              j = ll + ll
              do while(j .LE. ir)
                  if(j.lt.ir)then
                      if(ra(j).lt.ra(j+1))then
                          j = j + 1
```

CHAPTER 12. HOW WELL DOES HEAP SORT WORK?

```
                end if
             end if
             if(rra.lt.ra(j))then
                ra(i) = ra(j)
                i = j
                j = j + j
             else
                j = ir + 1
             end if
          end do
          ra(i) = rra
       end do
30     continue
       nin = 12
       open(nin, file=fou)
       do 230 i = 1, narr
          rra = ra(i)
          le = LEN_TRIM(rra(1:lenrec))
          write(nin,'(a)')rra(1:le)
230    continue
       close(nin)
       return
       end
```

The test data is a file 500,000,000 bytes long, which contains 1,000,000 records each containing 500 bytes, where each byte, except the end-of-line byte, is an upper-case letter selected randomly so all are equally likely. The time to execute the procedure that sorts this input data located in an existing file and writes the sorted records to a new file is about 13 seconds with the hardware and software used here. The conclusion is that the heap sort subroutine used here will likely adequately handle one year of water quality assessment on a single frugal salvage laptop, for any state, and more capable computers exist that will likely be adequate for one year of data consolidated for all contiguous 48 states at one time. 13 seconds may be compared with two years, which is the time given states to collect, analyze, and submit water quality assessment reports.

Rationalizing why this took 13 seconds, we first notice that the data are essentially random, and, one record contains 499 bytes, and, there are 1,000,000 records. The key step in the sort is comparing two records, finding which record evaluates as greater than the other. For this compare,

CHAPTER 12. HOW WELL DOES HEAP SORT WORK?

because the data are random, only a small number of bytes at the beginning of the record needs to be compared, the entire tail-end of the record is never tested. So, we conclude that it would be misleading to claim that all files with 500 million bytes could be sorted in 13 seconds. Having more records with the same head-end will force the computer to search most of each record, which will greatly increase the time to sort the file. Writing specifications for constructing files that will be sorted should take this behavior into consideration, which is why it is often much better to have full control of all aspects of a computer programming project and have full knowledge of how even the insignificant parts impact the performance of the resulting product. Adding performance criteria to specifications is wise.

The coding is short, everything is in our control, so, why not edit a couple of lines and try out an idea, instead of continuously mulling it over in our minds. So rising to the challenge, all but the last 20 bytes in every record were overwritten with the same byte, and the example code for this chapter was rerun. The result: with completely random bytes in each record the run time was about 13 seconds, with the records having the random bytes only in the last 20 places took about 15 seconds. Conclusion: our mulling, above, now has some evidence for veracity. Was this a waste of time? No, this is an example of working with computers. It doesn't matter if what we are mulling is well known to professionals skilled in their craft, and has been published in many text books, it is all about our thinking and our mulling about an idea we have. The payoff is when we mull over something that is not well known to professionals, and, we hit on some idea, and we try it out, and it works. If we are computer literate, this happening is more likely than if we remain computer illiterate in which case it won't happen to us. It is the same as talking and listening, and, writing and reading. Exchange of ideas will more likely lead to useful ideas than if we all have minimal communication with each other. Yes, a solo coder and a computer are talking and listening to each other.

Putting this in the context of thinking, what do we have now in this computer code example of heap sort? We have a simple illustration of how a frugal laptop and free operating system and free compiler and the time and energy it takes to type in a page or two of computer code, can take a relatively huge amount of information, 500 million bytes worth of information, and put it in alphabetical order. That sounds simple but the implications are enormous. We can organize chaos, and once organized we can do something useful with it. This illustration is just one of infinite possible ways we can do something useful that in recent history was impossible. People started talking hundreds of thousand years ago or more, then reading and writing about three thousand years ago or more, now the beginning of computer technology for tens of years. What happens to society if some say, it's fine if you don't learn to read and write, you can be useful and productive in our society. What happens to society if some say, it's fine if you don't learn to talk, you can be useful and productive in our society. On the other hand, just because you learn to talk does not mean that you will be a professional talker or listener, although there are people who do that. Just because you learn how to read and write does not mean you will be a professional writer or reader, although there are people who do that. Just because you become computer literate does not mean you will become a computer scientist or a professional coder, although there are people who do that. Talking and writing are essential basic literacy skills that modern people need so that they can become useful and productive in our society, and now computer literacy should be recognized by society as necessary for modern people to become useful and productive in our society, as necessary as talking and writing. There exists and has always existed uneven distributions of literacy, that is, talking skills and writing skills, in the world and the various social structures living in the world. The same is and will be true of computer literacy. That is human nature, but when some people encourage and facilitate the acquisition of literacy then all societies benefit and some of the benefits fall on the illiterate. That perhaps is all we can hope for. You now, with this example of what heap sort can do, would be called computer literate by many, especially if you first acquired a computer, installed an operating system, then entered the code yourself into your laptop, compiled and executed it, and examined the output and convinced yourself that it works as expected. No one handed it to you, you did it all by yourself. That is computer literacy, and infinite other possibilities are within grasp. Imagine what else can be done if five hundred million chunks of chaos can be organized, by you on your frugal laptop with free software and comparatively little time and effort, in a few seconds. Frugal does not mean inferior.

Chapter 13

Looking Inside Files

Often it is necessary for efficient progress, to look at the bytes in a file, usually to understand what is going wrong with a program that is being written. There are utilities that display the contents of files, however, here is the code to view a file, so that greater understanding about how a file is constructed and what is usually invisible to a view provided by a text editor. Files are little more than a sequence of bytes. There is more information about a file than is contained in the bytes inside the file, such as how long is the file, when it was created, et cetera. Also important when things are not going well is the file system, in which the file is located. Details of file systems is left to the reader to explore. A taste, the file system type used by Crunchbang Linux is named ext4, which may be found by going to Terminal and entering on the keyboard df -T followed by the Enter-key. In the middle-ish of what splashes on the black screen of Terminal will be ext4. The rest of the stuff on the screen is meaningful only to Linux geeks.

After creating the source code file with the Text Editor, in Terminal enter

gfortran C_13_SeeFileBytes.f90

and a file named a.out will appear in the same directory. To view a file with, say the name, untitled.txt, in Terminal enter

./a.out untitled.txt

and press the Enter-key. The first 200 bytes of the file will be displayed. How to display any block of 200 bytes in the file is evident on the screen.

```
! this file name:    C_13_SeeFileBytes.f90
```

```
! dependencies: none
! to compile:
!          gfortran -o D_13_run.out
!                          C_13_SeeFileBytes.f90
! to execute:  ./D_13_runseefile.out
!                          somefilename.xxx
! This example code illustrates:
!         see the bytes in a file
! NOTE: this program is interactive,
!       and needs file name
!       in the command line,
!       see code below for details
       program SeeFileBytes

! for the input file-name from the argument
       character*60 arg
       character*256 fin

! to have the file in memory
       integer(1), allocatable :: narr(:)
       integer(1) ib

! the line that will be written
       integer(1) nscribe
       dimension nscribe(100)
       character(100) cscribe
       equivalence(nscribe,cscribe)

! get the filename from the call arguments
       call get_command_argument(1,arg)
       fin = TRIM(arg)

! get size of the file in bytes
       write(*,*)'file name = ',fin(1:20)

       INQUIRE(FILE=fin,SIZE=lenfile)
       write(*,*)'filesize =',lenfile

! import the file into narr()
       allocate(narr(lenfile))
       open(UNIT=7,FILE=fin,ACCESS='DIRECT', &
                              RECL=lenfile)
```

CHAPTER 13. LOOKING INSIDE FILES

```
            read(7,REC=1)narr
            close(7)

!  parameters for the display of the bytes
            nrows = 10
            ncols = 10
            nsize = nrows * ncols

!  display this character instead of
!  a non-printing character
            nsted = 46

!  index for a byte in array narr(),
!  starting point for display
            num = 1

100         continue

!  loop display over rows
            write(*,*)'filesize = ',lenfile
            now = num
            do 90 jrow = 1, nrows
                nscribe = 32              ! blank the line
                write(cscribe(1:8),'(i5)')num
                do 80 jcol = 1, ncols
                    ib = narr(num)
                    num = num + 1
                    ii = 10 + 4 * jcol
                    write(cscribe(ii:ii+3),'(i4)')ib
                    if(ib.LT.32.OR.ib.GT.126)ib = nsted
                    jj = 10 + 45 + jcol
                    nscribe(jj) = ib
80              continue

                write(*,*)cscribe
90          continue

!  location of the next group of bytes to display
            num = now + 100
            write(*,*)'current line position =',num, &
                    '  enter new current position:'
            read(*,'(i15)',IOSTAT=itry,   &
```

```
                                    advance='no') next
        if(itry  .NE.  −2)go to 200
        if(next.GT.0)num = next
        if(num.GE.0)go to 100

200     continue

        deallocate(narr)

        end
```

This utility is useful for examining every byte at every location in a file. A useful view of some details important to coding can be obtained in Geany, the Crunchbang Text Editor. Select View from Geany's top toolbar, then select Editor, then select as needed, Show White Space, Show Line Endings, Show Indentation Guides.

Chapter 14

Merge

The last tool to be placed on the workbench is merge. This tool takes two files containing data in the form of records, exactly as used to demonstrate heap sort. These two files are first sorted, and then combined into one file which is also sorted, but sorted in the merge process.

The following code generates test data in two files, sorts each file, then merges the two files into one file. Examining the input and the output shows that the computer programs work as intended.

```python
# this file name:   C_14_merge_doit.py
# dependencies:   C_14_mergedata.f90
#                 C_14_merge.f90
# to execute:   python C_14_merge_doit.py
# This example code illustrates:
#              using Python to automate tasks
import os
os.system \
  ('gfortran -o D_14_data.out C_14_mergedata.f90')
os.system('./D_14_data.out')
os.system \
  ('gfortran -o D_14_run.out C_14_merge.f90')
os.system('./D_14_run.out')
```

```
! this file name:   C_14_mergedata.f90
! dependencies: none
! to compile:
!       gfortran -o D_14_data.out
!                   C_14_mergedata.f90
! to execute:   ./D_14_data.out
! This example code illustrates:
!           generate simulated data
      program C_14_mergedata

      character(256) fou

      numlines = 999993
      lengline = 50                  ! includes LF
      iseed = 1357
      fou = 'E_14_mergedataONE.txt'
      call getdata(iseed,numlines,lengline,fou)

      numlines = 934567
      lengline = 50                  ! includes LF
      iseed = 7531
      fou = 'E_14_mergedataTWO.txt'
      call getdata(iseed,numlines,lengline,fou)

      end
!_____
      subroutine getdata(iseed,numlines, &
                              lengline,fou)
      character(256) fou
      integer(1), allocatable :: narr(:)

      lengfile = lengline * numlines
      allocate(narr(lengfile))
      narr = 0

! initialize pseudorandom number generator
      x = RAND(iseed)

! fill with ASCII readable data
! plus LineFeed at end of record
      do 30 j = 1, numlines
```

CHAPTER 14. MERGE

```fortran
                nstart = lengline * (j - 1) + 1
                nstop  = nstart + lengline - 2
                do 20 k = nstart, nstop
                   ! upper case letters A through Z
                   narr(k) = INT(RAND() * 26.0) + 65
20              continue
                narr(nstop+1) = 10

30      continue

        inquire(IOLENGTH=lengnarr) narr
        open(UNIT=7,FILE=fou,ACCESS='DIRECT',  &
                            RECL=lengnarr,    &
                            ACTION='READWRITE')
        write(7,REC=1) narr
        close(7)
        deallocate(narr)

        end subroutine
```

```fortran
! this file name:   C_14_merge.f90
! dependencies: none
! to compile:   gfortran -o D_14_run.out
!                        C_14_merge.f90
! to execute:   ./D_14_run.out
! This example code illustrates:
!          merge two sorted files
       program C_14_merge

       character*256 finA, finB, fou
       character cuf

       finA = 'E_14_SortedONE.txt'
       finB = 'E_14_SortedTWO.txt'
       fou  = 'E_14_mergedoutput.txt'

       call system( &
'sort E_14_mergedataONE.txt > E_14_SortedONE.txt')
       call system( &
'sort E_14_mergedataTWO.txt > E_14_SortedTWO.txt')
```

CHAPTER 14. MERGE

```
      INQUIRE(FILE=finA,SIZE=lenfileA)
      INQUIRE(FILE=finB,SIZE=lenfileB)

      nin = 11
      open(nin,file=finA)
      cuf = char(32)

      lenrec = 0
      do while (ichar(cuf) .NE. 10)
         call fgetc(nin,cuf,istat)
         if(istat.NE.0)go to 999
         lenrec = lenrec + 1
      end do
      close(nin)

      numrecA = lenfileA / lenrec
      numrecB = lenfileB / lenrec

      call mergeit(lenrec,numrecA,numrecB, &
                             finA,finB,fou)

999   continue

      end
```
!――――――――――――――――――――――――――――――――――――
```
      subroutine mergeit(lenrec,numrecA,numrecB, &
                            finA,finB,fou)

      character*256 finA, finB, fou

      character(lenrec), allocatable :: ra(:)
      character(lenrec), allocatable :: rb(:)

      allocate(ra(numrecA))
      allocate(rb(numrecB))

      nin = 11
      open(nin,file=finA)
      do j = 1, numrecA
         read(nin,'(a)')ra(j)
```

CHAPTER 14. MERGE

```
      end do
      close(nin)

      open(nin, file=finB)
      do j = 1, numrecB
         read(nin,'(a)')rb(j)
      end do
      close(nin)

      nou = 12
      open(nou, file=fou)

      ja = 1
      jb = 1
      do while( (ja <= numrecA) .AND.    &
                              (jb <= numrecB) )
         if( ra(ja) .LE. rb(jb) )then
            write(nou,'(a)')ra(ja)
            ja = ja + 1
         else
            write(nou,'(a)')rb(jb)
            jb = jb + 1
         end if
      end do

      do while( ja .LE. numrecA )
         write(nou,'(a)')ra(ja)
         ja = ja + 1
      end do

      do while( jb .LE. numrecB )
         write(nou,'(a)')rb(jb)
         jb = jb + 1
      end do

      close(nin)

      deallocate(ra)
      deallocate(rb)

      return
      end
```

Chapter 15

Imperfect Data and Bayes Theorem

Almost no observational data are perfect, that is, free from errors of every type possible, which means that the data do not truly represent the true behavior of what is being measured. Bayes Theorem presumes that all the data are perfect. The data, which are both the input and the output of Bayes Theorem, are numbers that obey the axioms of probability, and, the axioms of probability define the perfect behavior of numeric quantities named probabilities. Perfect behavior includes the numbers put in and the numbers that come out of Bayes Theorem, and includes what happens to those numbers while inside Bayes Theorem, which for us is simply a bunch of symbols that all together look like an algebraic equation. As discussed above, Bayes Theorem works perfectly, so the output is perfect if and only if the input is perfect. The input, for our use of Bayes Theorem as a tool to assist our process of assessment, is observations and measurements of water and measurables related to water. There is no perfect measurement of some quality or characteristic of a real substance, by definition of "measurement", which requires a "ruler" which does not have infinitesimal marks separating and distinguishing one "distance" from another. Not only is the measurement imperfect, but the real substance is imperfect. Take a glass of water, with ice cubes. Measure density anywhere inside the glass, multiple times. At every point measured is water, but where you measure in this glass of water will impact your results, as liquid water has a different density than solid water.

When imperfect data is put into Bayes Theorem, imperfect data comes out. Bayes Theorem presumes, actually requires, that the input data is perfect. The imperfect data coming out of Bayes Theorem may not agree

CHAPTER 15. IMPERFECT DATA AND BAYES THEOREM

with common sense. We have a problem.

Look at what is used as input, and what is used as output. Numbers are used. The numbers represent probabilities. Probabilities, by definition, range in value from zero, meaning impossible, to one, meaning certain. No other numbers are allowed, by definition. Define a new probability, name it "representativeness", require that "representativeness" obey all the axioms of probability, but defer looking deeper.

Next, look at what happens with Bayes Theorem processing data. If the input data is totally imperfect then we should ignore what comes out of Bayes Theorem. If the input data is totally perfect, then we should completely believe what comes out of Bayes Theorem.

We use Bayes Theorem as an update of our belief given input data. If we ignore what comes out of Bayes Theorem, we end up keeping our prior belief unchanged. If we completely believe what comes out of Bayes Theorem, we end up replacing our prior belief with this current belief. In between these two extremes, what comes out should be something between our prior belief and the current result coming out of Bayes Theorem. This can be written algebraically two ways,

where "new" = N = the probability we carry forward,
"old" = P = our prior probability,
"result" = R = probability coming out of Bayes Theorem,
"representativeness" = B = number between zero and one,
then
$N = R * B + P * (1 - B)$
which reads as "our new probability that we carry forward
equals
the result of Bayes Theorem
times
the representativeness
plus
our probability that we carried forward
times
.NOT. the representativeness
meaning adding a fraction of the previous probability
to a fraction of the resulting probability
giving the new probability
such that the two fractions add to one,
or,
as a second equivalent expression,
$N = P + (R - P) * B$
which looks familiar to someone familiar with vectors,
meaning, take a starting point and

CHAPTER 15. IMPERFECT DATA AND BAYES THEOREM

add
the distance between the start and end points
times
a fraction of the distance between the start and end.

In the above, a number called "representativeness" is introduced and required to range from zero to one. This number is defined to behave as the probability that the measured value of a characteristic is truly representative of the water being measured, which encompasses errors in sampling, errors in location, errors in time, errors in measurement method, errors committed by the person measuring, errors in the measurement equipment, and any other errors the devil's advocate wishes to call to our attention. If the errors are recognized and fixable, then fix it, instead of using representativeness. Let representativeness be used only for issues that are truly not fixable.

As a probability, we can use Bayes Theorem to update the value of this probability called representativeness as new data are processed.

Chapter 16

Probability Demo

Probability studies events that occur many times, and focuses on the likelihood that a particular event has a particular outcome. Many textbooks use concrete examples for events, such as coins that are flipped, dice that are rolled, a deck from which cards are drawn, two urns containing blue and red marbles from which a marble is withdrawn. A coin is flipped, which is an event, and the outcome is heads or tails, and probability can provide some insight on the likelihood that a coin flip will result with heads.

With a computer, an event is simulated, and outcomes are generated. An event is anything a computer can simulate, and an outcome result is anything a computer simulation can generate. The computer simulation and generated outcomes are presumed to be very carefully thought through and truly represent what is being simulated. The hope is that most of the time, the simulation of events happening in the real world reasonably represents what is being simulated.

In this demo that illustrates various relationships of probability, each demo starts with a Fortran provided function that generates pseudo-random numbers. A truly random number does not depend on anything else that happens when that number is produced. In the jargon of mathematics, these numbers are called "independent" of each other, that is, a particular number does not depend on anything that has happened before, or after, or anything else happening at the same time. Our laptop PC all by itself, that is, no interaction with the world outside the laptop, is not capable of producing truly random numbers, it can produce pseudo-random numbers which for the purposes of the demonstrations in this chapter, are close enough to truly random numbers.

More about "event", an event is something happening, usually something in the real world does something. For example, a coin is something

CHAPTER 16. PROBABILITY DEMO

in the real world. A coin flip is an event happening to the coin. "Heads" is one outcome of the event, one outcome of a flip. All this is in the real world and is observable or measurable, and we end up with a description for something qualitative and a number for something quantitative. "Heads" is a qualitative observation.

The demos in this chapter simulate reality by creating a sequence of qualitative outcomes. This is done by first generating a pseudo-random number, usually one pseudo-random number for each event, and then converting that number into a qualitative idea, "true" and "false". We introduce another number and call it "threshold". In the demos here, there are two thresholds, one for each of two simultaneous events.

To convert pseudo-random numbers into either "true" or "false", compare the pseudo-random number with the threshold, and say "if the random number is less than the threshold then something is true, otherwise that something is false". For now, we don't care what "that something" is, only that it is true or false. Later we will take this abstract something and put in a concrete idea, such as something related to water quality. But for now we can leave it unspecified, which is good, for then we can put in anything that is really interesting.

Next step is counting the number of "true" and the number of "false" outcomes for each "event of something",

Finally, divide the total number of "true" by the total number of events, both assigned to a "something", which gives a reasonable estimate of the probability that an event will yield a "true" for that "something". The pickiness with words here reflect that in these demos there are more than one "something", more than one "event", more than one "threshold", more than one "outcome", and, we have to keep straight what is associated with what and what order things happen, or the details quickly get garbled in our heads.

A summary of our process in these demos, 1) thresholds for each independent event are established, 2) the number of trials is established, 3) for each trial, for each event, a pseudo-random number is produced, 4) each random number is compared with the threshold for that event and the result is either "true" or "false", 5) for each event, the number of "true" is counted or the number of "false" is counted, 6) at the end of the trials, the total number of "true" is divided by the number of trials, yielding another number that ranges from zero to one, which we accept as a reasonable approximation of "the probability that a specified event will be true". We can do the same for "false".

CHAPTER 16. PROBABILITY DEMO

```python
# this file name:   C_16_probability_doit.py
# dependencies:   C_16_think_probability.f90
#                 C_16_coinflip.f90
# to execute:   python C_16_probability_doit.py
# This example code illustrates:
#           using Python to automate tasks
import os
os.system( \
    'gfortran -o D_16_rb.t C_16_think_prob.f90 ')
os.system('./D_16_rb.t > E_16_see_prob.txt ')
os.system( \
    ' gfortran -o D_16_rn.t C_16_coinflip.f90 ')
os.system('./D_16_rn.t > E_16_see_coinflip.txt ')
```

```fortran
! this file name:   C_16_think_prob.f90
! dependencies: none
! to compile:
!    gfortran -o D_16_run.out C_16_think_prob.f90
! to execute:   ./D_16_run.out
! to execute and save results:
!          ./D_16_run.out > E_see_16_results.txt
! This example code illustrates:
!    some concepts of probability, simulated
!  program C_16_think_probability
!   demonstrations of several probability
!   relationships using two independent real
!   numbers, each representing an event
    write(*,*)' A summary of our process in these'
    write(*,*)' demos, 1) thresholds for each '
    write(*,*)' independent event are established,'
    write(*,*)' 2) the number of trials is'
    write(*,*)' established, 3) for each trial,'
    write(*,*)' for each event, a pseudo-random'
    write(*,*)' number is produced,4) each random'
    write(*,*)' number is compared with the '
    write(*,*)' threshold for that event and the'
    write(*,*)' result is either "true" or'
    write(*,*)' "false", 5) for each event, the'
    write(*,*)' number of "true" is counted'
    write(*,*)' or the number of "false" is'
```

CHAPTER 16. PROBABILITY DEMO

```
          write(*,*)' counted, 6) at the end of the'
          write(*,*)' trials, the total number of "true"'
          write(*,*)' is divided by the number of trials'
          write(*,*)' yielding another number that'
          write(*,*)' ranges from zero to one, and we'
          write(*,*)' accept as a reasonable'
          write(*,*)' approximation of "the probability'
          write(*,*)' that the event is true".'
! range is from zero to rnge
          rnge = 1.0
! number of trial threshold values
          nthresh = 3
! fraction of range for first threshold value
          thresh = 1.0 / REAL(nthresh + 1)
! Loop over threshold values
          write(*,*)'ooooo BEGIN THIS ILLUSTRATION ooooo'
          do 301 jta = 1, nthresh
             do 300 jtb = 1, nthresh
! (the end of these loops is at the
!  very end of this program)
! the numeric value of each
! threshold for this trial
                thresholdA = REAL(jta) * thresh * rnge
                thresholdB = REAL(jtb) * thresh * rnge
! Begin the demos at these threshold values.
          write(*,*)
          write(*,*)'ooooooooooooooooooooooooooooooooooo'
          write(*,*)'--- this loop has eleven demos---'
          write(*,*)'--- the loop parameters are: ---'
          write(*,*)' jta,jtb,thresholdA,thresholdB = ', &
                        jta,jtb,thresholdA,thresholdB
          write(*,*)'----------------------------------'
! ======== FIRST DEMO =========================
!        simple P( A ) and simple P( NA )
          massive = 100000
          iseed = 123579
          x = RAND(iseed)
          numbtrueA = 0
          numbtrueNA = 0
          do 201 j = 1, massive
             valueA = RAND()
             valueB = RAND()
```

CHAPTER 16. PROBABILITY DEMO

```
         if (valueA .GT. thresholdA) numbtrueA =   &
                                       numbtrueA + 1
         if (valueA .LE. thresholdA) numbtrueNA =   &
                                       numbtrueNA + 1
201    continue
       probtrueA  = REAL(numbtrueA)  / REAL(massive)
       probtrueNA = REAL(numbtrueNA) / REAL(massive)
       write(*,*)
       write(*,*) 'DEMO 1 ============================='
       write(*,*) '   simple P( A ) and simple P( NA )'
       write(*,*) '          thresholdA =', thresholdA
       write(*,*) '          probtrueA, probtrueNA =',  &
                            probtrueA, probtrueNA
       write(*,*) ' Compare these two numbers'
       write(*,*) ' with the value of threshold.'
       write(*,*)'_____'
       write(*,*)
!  ======== SECOND DEMO ==================
!         P( A .AND. B ) = P(A) * P(B)
       massive = 1000000
       iseed = 123579
       x = RAND(iseed)
       numbtrueAaB = 0
       numbtrueA = 0
       numbtrueB = 0
       do 202 j = 1, massive
          valueA = RAND()
          valueB = RAND()
          if (valueA .GT. thresholdA) numbtrueA =   &
                                       numbtrueA + 1
          if (valueB .GT. thresholdB) numbtrueB =   &
                                       numbtrueB + 1
          if (valueA .GT. thresholdA .AND.  &
              valueB .GT. thresholdB)       &
                        numbtrueAaB = numbtrueAaB + 1
202    continue
       probtrueA = REAL(numbtrueA) / REAL(massive)
       probtrueB = REAL(numbtrueB) / REAL(massive)
       productAB = probtrueA * probtrueB
       probtrueAaB =   &
                REAL(numbtrueAaB) / REAL(massive)
       calc = ( 1.0 - thresholdA) *   &
```

CHAPTER 16. PROBABILITY DEMO

```
                             (  1.0  -  thresholdB  )
    write(*,*)
    write(*,*) 'DEMO 2'
    write(*,*) '  == P( A .AND. B ) = P(A) * P(B) =='
    write(*,*) '     thresholdA, thresholdB =',  &
                     thresholdA, thresholdB
    write(*,*) '     probability both .GT. threshold,'
    write(*,*) '     that is, ( A .AND. B ) is .TRUE.'
    write(*,*) '       probtrueAaB =', probtrueAaB
    write(*,*) '  think: the intersection where both'
    write(*,*) '  A and B are .TRUE. which is'
    write(*,*) '  ( (1.0 - thresholdA) *'
    write(*,*) '                    (1.0 - thresholdB) )'
    write(*,*) '  in this simplistic demo only'
    write(*,*) '              calc =', calc
    write(*,*) '  probtrueA  =', probtrueA
    write(*,*) '  probtrueB  =', probtrueB
    write(*,*) '  productAB  =', productAB
    write(*,*) '     which should equal P(A .AND. B)'
    write(*,*)'_____'
    write(*,*)
!=========== THIRD DEMO =========================
!    P( S ) = P( A ) + P( NA )
    massive = 1000000
    iseed = 123579
    x = RAND(iseed)
    numbtrueA = 0
    numbtrueNA = 0
    do 203 j = 1, massive
       valueA = RAND()
       valueB = RAND()
       if(valueA.GT.thresholdA)numbtrueA =  &
                               numbtrueA + 1
       if(valueA.LE.thresholdA)numbtrueNA =  &
                               numbtrueNA + 1
203    continue
    probtrueA = REAL(numbtrueA) / REAL(massive)
    probtrueNA = REAL(numbtrueNA) / REAL(massive)
    probtotal = probtrueA + probtrueNA
    write(*,*)
    write(*,*) 'DEMO 3'
    write(*,*) '  ===  P( S ) = P( A ) + P( NA )  ==='
```

CHAPTER 16. PROBABILITY DEMO

```
      write(*,*)'              thresholdA =',thresholdA
      write(*,*)'              probtotal  =',probtotal
      write(*,*)'                          Should equal 1'
      write(*,*)' This is a probability axiom,'
      write(*,*)' which means, it is obvious,'
      write(*,*)' or should be obvious.'
      write(*,*)'---------------------------------------'
      write(*,*)
!=========== FOURTH DEMO =================
!     the general addition rule
!  P( A .OR. B ) = P( A ) + P( B ) - P( A .AND. B )
      massive = 1000000
      iseed = 123579
      x = RAND(iseed)
      numbtrueA = 0
      numbtrueB = 0
      numbtrueAB = 0
      numbtrueC = 0
      do 204 j = 1, massive
         valueA = RAND()
         valueB = RAND()
         if(valueA.GT.thresholdA)numbtrueA = &
                                 numbtrueA + 1
         if(valueB.GT.thresholdB)numbtrueB = &
                                 numbtrueB + 1
         if(valueA.GT.thresholdA .AND.     &
            valueB.GT.thresholdB)          &
                     numbtrueAB = numbtrueAB + 1
         if(valueA.GT.thresholdA .OR.      &
            valueB.GT.thresholdB)          &
                     numbtrueC = numbtrueC + 1
204   continue
      probtrueA = REAL(numbtrueA) / REAL(massive)
      probtrueB = REAL(numbtrueB) / REAL(massive)
      probtrueAB = REAL(numbtrueAB) / REAL(massive)
      probtrueC = REAL(numbtrueC) / REAL(massive)
      calcC = probtrueA + probtrueB - probtrueAB
      write(*,*)
      write(*,*)'DEMO 4  =========================='
      write(*,*)' P( A .OR. B ) = P( A ) + P( B ) -'
      write(*,*)'                 P( A .AND. B )'
      write(*,*)' thresholdA, thresholdB =', &
```

CHAPTER 16. PROBABILITY DEMO

```
                       thresholdA , thresholdB
     write(*,*)'   probtrueA , probtrueB , probtrueAB  =', &
                       probtrueA ,  probtrueB , probtrueAB
     write(*,*)'   calcC , probtrueC =',calcC , probtrueC
     write(*,*)'        This is called the General'
     write(*,*)'        Addition Rule.'
     write(*,*)'        Nice demo, important rule,'
     write(*,*)'        but not used here.'
     write(*,*)'------------------------------------------'
     write(*,*)
!=========== FIFTH DEMO =======================
!   P( A | B ), conditional probability
     massive = 1000000
     iseed = 123579
     x = RAND(iseed)
     numbtrue = 0
     numbtest = 0
     do 205 j = 1, massive
       valueA = RAND()
       valueB = RAND()
       if(valueB .GT. thresholdB) then
         numbtest = numbtest + 1
         if(valueA .GT. thresholdA) numbtrue = &
                                    numbtrue + 1
       end if
205  continue
     probtrue = REAL(numbtrue) / REAL(numbtest)
     write(*,*)
     write(*,*) 'DEMO 5 ==== P( A | B ) ============'
     write(*,*)'   thresholdA , thresholdB =', &
                       thresholdA , thresholdB
     write(*,*)'   probability ( A | B ) is .TRUE. =',&
                                              probtrue
     write(*,*)'        This is called'
     write(*,*)'        conditional probability,'
     write(*,*)'        and is the quantity that'
     write(*,*)'        Bayes Theorem calculates.'
     write(*,*)'------------------------------------------'
     write(*,*)
!=========== SIXTH DEMO =======================
!   P( A | B ) = P( A .AND. B ) / P( B )
     massive = 1000000
```

```
      iseed = 123579
      x = RAND( iseed )
      numbtrueB = 0
      numbtrueAaB = 0
      numbtrue = 0
      numbtest = 0
      do 206 j = 1, massive
        valueA = RAND()
        valueB = RAND()
        if ( valueB .GT. thresholdB ) numbtrueB =   &
                                      numbtrueB + 1
        if ( valueA .GT. thresholdA .AND.   &
            valueB .GT. thresholdB )   &
                         numbtrueAaB = numbtrueAaB + 1
        if ( valueB .GT. thresholdB ) then
          numbtest = numbtest + 1
          if ( valueA .GT. thresholdA ) numbtrue =   &
                                        numbtrue + 1
        end if
206   continue
      probtrueB = REAL( numbtrueB ) / REAL( massive )
      probtrueAaB = REAL( numbtrueAaB ) /   &
                                    REAL( massive )
      pcalc = probtrueAaB / probtrueB
      probtrueAgivenB = REAL( numbtrue ) /   &
                                    REAL( numbtest )
   write(*,*)
   write(*,*) 'DEMO 6 ============================='
   write(*,*) '    P( A | B ) ='
   write(*,*) '              P( A .AND. B ) / P( B )'
   write(*,*) '    thresholdA, thresholdB =',   &
                  thresholdA, thresholdB
   write(*,*) '    P( A | B ) calculated directly,'
   write(*,*) '            probtrueAgivenB =',   &
                                probtrueAgivenB
   write(*,*) ' the same calculated from the parts,'
   write(*,*) '                    pcalc =', pcalc
   write(*,*) ' This is one way to'
   write(*,*) ' write Bayes Theorem.'
   write(*,*)'---------------------------------------'
   write(*,*)
!=========== SEVENTH DEMO =====================
```

CHAPTER 16. PROBABILITY DEMO

```
!      P( A ) = 1.0 - P( .NOT. A )
       massive = 1000000
       iseed = 123579
       x = RAND(iseed)
       numbtrueA = 0
       numbtrueNA = 0
       do 207 j = 1, massive
         valueA = RAND()
         valueB = RAND()
         if(valueA.GT.thresholdA)numbtrueA = &
                                        numbtrueA + 1
         if(valueA.LE.thresholdA)numbtrueNA = &
                                        numbtrueNA + 1
207    continue
       probtrueA = REAL(numbtrueA) / REAL(massive)
       probtrueNA = REAL(numbtrueNA) / REAL(massive)
       calcA = 1.0 - probtrueNA
      write(*,*)
      write(*,*) 'DEMO 7'
      write(*,*)'  === P( A ) = 1.0 - P( .NOT. A ) ==='
      write(*,*)'     thresholdA =',thresholdA
      write(*,*)'     probtrueA, calcA =',probtrueA, calcA
      write(*,*)'                        should be the same'
      write(*,*)'    This is a term in Bayes Theorem.'
      write(*,*)'_____'
      write(*,*)
!========== EIGHTH DEMO =========================
! P( B | A ) = 1.0 - P( B | .NOT. A )
       massive = 1000000
       iseed = 123579
       x = RAND(iseed)
       numbtrueBgA = 0
       numbtestBgA = 0
       numbtrueBgNA = 0
       numbtestBgNA = 0
       do 208 j = 1, massive
         valueA = RAND()
         valueB = RAND()
         if(valueB.GT.thresholdB)then
           numbtestBgA = numbtestBgA + 1
           if(valueA.GT.thresholdA)numbtrueBgA = &
                                        numbtrueBgA + 1
```

```
              end if
              if ( valueB . GT. thresholdB ) then
                 numbtestBgNA = numbtestBgNA + 1
                 if ( valueA . LE . thresholdA )numbtrueBgNA  =  &
                                              numbtrueBgNA + 1
              end if
208        continue
           probtrueBgA = REAL(numbtrueBgA) / &
                                         REAL( numbtestBgA )
           probtrueBgNA = REAL(numbtrueBgNA) /    &
                                         REAL( numbtestBgNA )
           calcBgA = 1.0 − probtrueBgNA
        write ( * , * )
        write ( * , * ) 'DEMO 8 ============================='
        write ( * , * ) ' P( B | A ) ='
        write ( * , * ) '              1.0 − P( B | .NOT. A )'
        write ( * , * ) ' thresholdA =', thresholdA
        write ( * , * ) ' probtrueBgA , probtrueBgNA , calcBgA =',&
                         probtrueBgA , probtrueBgNA , calcBgA
        write ( * , * ) ' This is another term'
        write ( * , * ) ' in Bayes Theorem.'
        write(*,*)'————————————————————————————————————'
        write ( * , * )
!=========== NINTH DEMO ======================
! P( A | B ) = P( B | A ) * P( A ) / P( B )
           massive = 1000000
           iseed = 123579
           x = RAND( iseed )
           numbtrueA = 0
           numbtrueB = 0
           numbtrueAgB = 0
           numbtestAgB = 0
           numbtrueBgA = 0
           numbtestBgA = 0
           do 209 j = 1, massive
              valueA = RAND()
              valueB = RAND()
              if ( valueA . GT. thresholdA ) numbtrueA =  &
                                            numbtrueA + 1
              if ( valueB . GT. thresholdB ) numbtrueB =  &
                                            numbtrueB + 1
              if ( valueB . GT. threshold ) then
```

CHAPTER 16. PROBABILITY DEMO

```
              numbtestAgB = numbtestAgB + 1
              if (valueA.GT.thresholdA)numbtrueAgB = &
                                    numbtrueAgB + 1
           end if
           if (valueA.GT.threshold) then
              numbtestBgA = numbtestBgA + 1
              if (valueB.GT.thresholdB)numbtrueBgA = &
                                    numbtrueBgA + 1
           end if
  209   continue
        probtrueA = REAL(numbtrueA) / REAL(massive)
        probtrueB = REAL(numbtrueB) / REAL(massive)
        probtrueAgB = REAL(numbtrueAgB) / &
                                    REAL(numbtestAgB)
        probtrueBgA = REAL(numbtrueBgA) / &
                                    REAL(numbtestBgA)
        calcAgB = probtrueBgA * probtrueA / probtrueB
       write(*,*)
       write(*,*) 'DEMO 9 ============================'
       write(*,*)' P( A | B ) ='
       write(*,*)'        P( B | A ) * P( A ) / P( B )'
       write(*,*)'   thresholdA, thresholdB =', &
                   thresholdA, thresholdB
       write(*,*)'   probtrueBgA, probtrueA, probtrueB='
       write(*,*)'  ', probtrueBgA, probtrueA, probtrueB
       write(*,*)'   probtrueAgB, calcAgB=', &
                   probtrueAgB, calcAgB
       write(*,*)'   This is another way'
       write(*,*)'   to write Bayes Theorem.'
       write(*,*)'----------------------------------------'
       write(*,*)
!=========== TENTH DEMO =================
! P( B ) = P( B | A ) * P( A ) +
!          P( B | .NOT. A ) * P( .NOT. A )
        massive = 1000000
        iseed = 123579
        x = RAND(iseed)
        numbtrueA = 0
        numbtrueNA = 0
        numbtrueB = 0
        numbtrueBgA = 0
        numbtestBgA = 0
```

```
      numbtrueBgNA = 0
      numbtestBgNA = 0
      do 210 j = 1, massive
        valueA = RAND()
        valueB = RAND()
        if(valueA.GT.thresholdA)numbtrueA = &
                                      numbtrueA + 1
        if(valueA.LE.thresholdA)numbtrueNA = &
                                      numbtrueNA + 1
        if(valueB.GT.thresholdB)numbtrueB = &
                                      numbtrueB + 1
        if(valueA.GT.thresholdA)then
          numbtestBgA = numbtestBgA + 1
          if(valueB.GT.thresholdB)numbtrueBgA = &
                                      numbtrueBgA + 1
        end if
        if(valueA.LE.thresholdA)then
          numbtestBgNA = numbtestBgNA + 1
          if(valueB.GT.thresholdB)numbtrueBgNA = &
                                      numbtrueBgNA + 1
        end if
210   continue
      probtrueA = REAL(numbtrueA) / REAL(massive)
      probtrueNA = REAL(numbtrueNA) / REAL(massive)
      probtrueB = REAL(numbtrueB) / REAL(massive)
      probtrueBgA = REAL(numbtrueBgA) / &
                                 REAL(numbtestBgA)
      probtrueBgNA = REAL(numbtrueBgNA) / &
                                 REAL(numbtestBgNA)
      calcB = probtrueBgA * probtrueA + &
              probtrueBgNA * probtrueNA
write(*,*)
write(*,*) 'DEMO 10 ==========================='
write(*,*)' P( B ) = P( B | A ) * P( A ) +'
write(*,*)'          P( B | .NOT. A ) * P( .NOT. A )'
write(*,*)'   thresholdA, thresholdB =', &
                thresholdA, thresholdB
write(*,*)'   probtrueA, probtrueNA =', &
                probtrueA, probtrueNA
write(*,*)'   probtrueB =', probtrueB
write(*,*)'   probtrueBgA, probtrueBgNA =', &
                probtrueBgA, probtrueBgNA
```

CHAPTER 16. PROBABILITY DEMO

```
      write(*,*)'   probtrueB,calcB=',probtrueB,calcB
      write(*,*)' This is the denominator in'
      write(*,*)' the previous expression.'
      write(*,*)'-------------------------------------------'
      write(*,*)
!=========== ELEVENTH DEMO =================
!    P( A | B ) = P( B | A ) * P( A ) /
!                  ( P( B | A ) * P( A ) +
!                    P( B | .NOT. A ) * P( .NOT. A ) )
      massive = 1000000
      iseed = 123579
      x = RAND(iseed)
      numbtrueA = 0
      numbtrueNA = 0
      numbtrueB = 0
      numbtrueBgA = 0
      numbtestBgA = 0
      numbtrueBgNA = 0
      numbtestBgNA = 0
      numbtrueAgB = 0
      numbtestAgB = 0
      do 211 j = 1, massive
        valueA = RAND()
        valueB = RAND()
        if(valueA.GT.thresholdA)numbtrueA = &
                                  numbtrueA + 1
        if(valueA.LE.thresholdA)numbtrueNA = &
                                  numbtrueNA + 1
        if(valueB.GT.thresholdB)numbtrueB = &
                                  numbtrueB + 1
        if(valueA.GT.thresholdA)then
          numbtestBgA = numbtestBgA + 1
          if(valueB.GT.thresholdB)numbtrueBgA = &
                                  numbtrueBgA + 1
        end if
        if(valueA.LE.thresholdA)then
          numbtestBgNA = numbtestBgNA + 1
          if(valueB.GT.thresholdB)numbtrueBgNA = &
                                  numbtrueBgNA + 1
        end if
        if(valueB.GT.thresholdB)then
          numbtestAgB = numbtestAgB + 1
```

```
              if ( valueA .GT. thresholdA ) numbtrueAgB  = &
                                          numbtrueAgB + 1
       end if
211    continue
       probtrueA      = REAL( numbtrueA )    / &
                                         REAL( massive )
       probtrueNA     = REAL( numbtrueNA )   / &
                                         REAL( massive )
       probtrueB      = REAL( numbtrueB )    / &
                                         REAL( massive )
       probtrueBgA    = REAL( numbtrueBgA )  / &
                                         REAL( numbtestBgA )
       probtrueBgNA   = REAL( numbtrueBgNA ) / &
                                         REAL( numbtestBgNA )
       probtrueAgB    = REAL( numbtrueAgB )  / &
                                         REAL( numbtestAgB )
       calcB = probtrueBgA * probtrueA + &
               probtrueBgNA * probtrueNA
       calcpost = probtrueBgA * probtrueA / calcB
       write ( * , * )
       write ( * , * ) 'DEMO 11 ============================='
       write ( * , * ) ' P( A | B ) = P( B | A ) * P( A ) /'
       write ( * , * ) '             ( P( B | A ) * P( A ) +'
       write ( * , * ) '    P( B | .NOT. A ) * P( .NOT. A ) )'
       write ( * , * ) '   thresholdA , thresholdB =' , &
                                    thresholdA , thresholdB
       write ( * , * ) '   probtrueA , probtrueNA , probtrueB =' ,&
                                    probtrueA , probtrueNA , probtrueB
       write ( * , * ) '   probtrueBgA , probtrueBgNA =' ,  &
                                    probtrueBgA , probtrueBgNA
       write ( * , * ) '      calcB =' , calcB
       write ( * , * ) '   probtrueAgB , calcpost =' , &
                                    probtrueAgB , calcpost
       write ( * , * ) ' This is the form of'
       write ( * , * ) ' Bayes Theorem used here.'
       write ( * , * ) '---------------------------------------'
       write ( * , * )
!== end of demonstrations for a threshold-pair ==
300    continue
301    continue
!======= end of loop over thresholds =========
       write ( * , * ) 'ooooooooooo----END----ooooooooooooooo'
```

end

```
! this file name:   C_16_coinflip.f90
! dependencies: none
! to compile:   gfortran -o D_16_runcoin.out
!                         C_16_coinflip.f90
! to execute:   ./D_16_runcoin.out
! This example code illustrates:
!   simulate flipping a coin
      program coin
      nheads = 0
      ntails = 0
      nflips = 0
      iseed = 98765
      x = RAND(iseed)
      do while (nflips .LT. 1000000)
          x = RAND()
          if( x .GT. 0.5 )then
              nheads = nheads + 1
          else
              ntails = ntails + 1
          end if
          nflips = nflips + 1
      end do
      write(*,*)nheads, ntails, nflips
      probheads = REAL(nheads) / REAL(nflips)
      write(*,*)probheads
      end
```

In this chapter, we experience computer technology being used to augment our thinking. If we want to understand probability, and have little or no prior exposure to ideas that deal with probability, traditionally we have two general approaches, one is analytic and the other is experimental.

Using analytical thinking we are shown the statements with the understanding that after we are shown the statements we will understand what the statements mean. Since this is difficult, there is another complementary and still traditional approach, we observe behaviors experimentally. An example is an arithmetic manipulative, that is, chips set out

CHAPTER 16. PROBABILITY DEMO   116

and counted, two piles of chips set up and counted to illustrate addition. For probability, the analytical approach takes us through a sequence of statements which are called the axioms and lemmas of probability, and for the experimental approach we are given a pair of dice, which we roll and tabulate the number of times the pair show 2, or 3, up to 12, and then we are suppose to connect our tabulations with the probability that a pair of dice will result in 2, and result in the other numbers through 12. In this chapter, we explore probability using both traditions, analytic and experimental. Each demo in the computer code above, illustrates either a definition, or an axiom, or a lemma of probability, in a sequence that is an outline of a derivation of Bayes Theorem. Each illustration is an experiment where an event such as rolling a pair of dice is performed by a computer program, not actually rolling dice, but representing what happens and what results when dice are rolled. Done manually with actual dice, you tabulate numbers. When a computer program runs, it tabulates numbers, doing the work of rolling the dice and tabulating the numbers. The difference between manual and computer is the number of experiments that can be performed in a given amount of time, and, the amount of data that can be tabulated, and the ease at which the experiment simulated on the computer may be tweaked or radically changed, enabling experiments completely out of reach for the manual approach. This partnership of human mind and computer allows a person to think both analytically and experimentally at the same time, not much different than before computers, except with computers the scope of the possible experiments has expanded enormously. The self-test for the thinker, after mulling the definitions, axioms, and lemmas, and studying their behaviors with computer simulation, is writing computer code where the ideas simulated in the computer behave identically with the ideas represented in the definitions, axioms, and lemmas. When successful, it indicates a level of competency and understanding, sufficient to apply the acquired skills and knowledge to problems, ideas, issues, tasks, in the real world.

In program coin, we test an axiom of probability, which is, given an event, then the sum of all independent probabilities of outcomes of the event equals one. Add together the probability of heads plus probability of tails, and see the number is very close to the number one, which agrees with the axiom. Program coin illustrates something simple enough to be obvious to many people, maybe requiring some mulling. Computers allow the human mind to create a simulation of something very complex or very large or both, something that the human mind finds very difficult to intuitively describe the behavior of this something. Making the computer run the simulation and produce the results, and then reviewing and mulling over the results, the human mind is able to mull through

CHAPTER 16. PROBABILITY DEMO

concepts that are un-mullable without computer assistance. An example, Bayes Theorem, Bayes mulled though it and his name is attached to it. Most people who have not studied probability do not mull and come up with Bayes Theorem. But, there are people who do. It is expected that many computer literate people who have not studied probability could mull with assistance from their computer, and come up with Bayes Theorem. This is the potential of the third greatest advance in human thinking.

Chapter 17

Translators

Introduced in the middle of Chapter 5, and buried in the above code that demonstrates probability relationships is the word translator. Repeating from Chapter 5, the probability of evidence B given impairment A, written symbolically is, $P(B \mid A)$, and which, as described later, is obtained by using a translator to convert data to probability.

Later is now. Some of the chapters, above, are about probability, starting with fundamental ideas and leading to Bayes Theorem, and attempts to facilitate some understanding of Bayes Theorem. Probability and Bayes Theorem are, for most people, difficult to understand, and time is needed for concepts to come together in the mind. And, the emphasis above is rephrasing the probability lingo into assessment lingo, which also for most people is difficult. So up to now, the translator has, for pedantic purposes, been left nebulous at best.

Clarity was the intent when all the pieces of an algebraic expression of Bayes Theorem were written out, the pieces separated by math symbols for add, subtract, multiply, divide, and equal. Clarity was the intent when each piece was described, in math representation, as a word, interpreted in assessment language. In all this is one piece that likely stood out to the astute reader, that is, $P(B \mid A)$, the probability of evidence B given impairment A. The astute reader likely thought, how is a numeric value assigned to this variable. All the other variables are easy, their values have obvious numeric values, more or less obvious. At least, the numeric values are easily found or obtained. But not $P(B \mid A)$, which is an enigma at this time in the development of the narrative of this book. It's taken a long effort to get here, but, for clarity purposes a lot of concepts needed to be introduced and explored before effort is spent on this enigma.

Diving in, we need a way that is understandable to the majority of citi-

CHAPTER 17. TRANSLATORS

zens who think about water quality assessment when we give this variable a numeric value. That is, when we are given an observation or measurement about some attribute of a water body, we convert that into a numeric value for the probability of evidence B given impairment A, where this numeric value ranges from zero to one, because it is a probability.

To do this, as an illustration, lets make a translator, say for a water body that is a source of drinking water, say for arsenic. On graph paper draw the two axis. Along the vertical axis put the numbers zero to one hundred, with zero at the origin, and label the vertical axis, Arsenic Concentration which means the result of the analysis of a water sample collected from the water body and expressed in concentration units of parts per billion, all strictly following the established protocols for sampling and analysis for this specific type of sample. Along the horizontal put some numbers with values between zero and one, with zero at the origin, and label the horizontal axis, Probability of Evidence given Impairment, which means, the horizontal axis is the probability that the numeric value of the corresponding observed concentration on the vertical axis will be the result of the procedures that follow the sampling and analysis protocols when the sample was taken from the water body and the water body is known to be impaired. It is important to remember that impaired is a binary concept while concentration is a continuum. We'll measure a concentration but we don't care at this time how much is actually in the water, we need to know only if the water body is impaired or not. In this simple situation, arsenic is both what is causing the impairment and what is observed, that is, measured directly. In other less simple situations, what is observed or measured may not be the same as what is causing the impairment. For translator development purposes, think for the vertical axis, these numbers are what I observe, one measurement at a time, and think for the horizontal axis, if the water body is impaired then this is the probability I will get this observed value, taken one observed value at a time.

We now have two axis on graph paper, and the label of each axis is explicitly and exactly the connection between the number we have and the number we want, that is, we have a numeric value for observed something and we want a numeric value for probability of that observed result occurring when the water body is impaired. Our enigma is now sketched out on graph paper.

Now the citizens, the civil servants, the bureaucrats, the lawyers, the stakeholders, the politicians, the experts, the experts hired by the stakeholders, the pundits, the distractors, the disruptors, the reporters, the commentators, the friends and neighbors, all gather around this graph paper. We need to draw a line on this graph paper so that given a numeric value of something observed, in this case arsenic concentration, we can look at the graph, go across until we reach a line, then go down to the horizontal axis and read off the numeric value of the probability that this arsenic concentration as observed will result if the water body is impaired. The graph is empty now except for the two axis, but, once the line is drawn on the graph paper, everything is easy.

Start out by everybody considering that the water body under consideration is impaired caused by arsenic. Go a little further, everybody think about imagining a situation where the water body is on the edge of impairment. On the edge means that you don't really know if it is impaired or not impaired. All you know is if the arsenic concentration is less, then it is not impaired, and if the concentration is more, then it is impaired. Do a lot listening to experts hired by the stakeholders and all the others who want to speak out, do a lot of arguing and hand waving, even do some deal making on the side, and, if power is at least slightly imbalanced between all the different viewpoints then this numeric value for the arsenic concentration observed when the water body is on the edge of impairment can be chosen, and enough people will more or less agree, and a few people will be happy. Assign for this numeric value of the observed concentration for a water body on the exact edge of impairment a probability of 50%. Think of a coin flip, the outcome is binary, either heads or tails equally for a fair coin. Before we flip the fair coin, we have no reason to believe that one outcome will occur more often than the other. The probability of a flip outcome is 50% heads, 50% tails. At the edge of impairment, we have no reason to believe that impairment is more or less likely than not-impaired. 50% probability for each outcome.

The key is that people make the decision, and, people use emotions, beliefs, financial gain or loss potential, preconceived notions, gut feelings, fake news, attitudes expressed on social media and public media, political propaganda, opinions expressed by science experts, logic, and, this list gets too long to be useful. The key is that science was not called upon to provide a definitive provable numeric value for this probability at the edge of impairment, but rather, people listened, studied, considered all the available information including information produced by science experts who used science and all the tools associated with science and came up with various studies including toxicity, which then the people may consider along with all the other information out there including fake science when choosing a numeric value of the observed concentration at the edge of impairment, and then the people make up their minds and choose a numeric value. There is no voice of authority, no dogma, and science with its objectivity is unable by itself to arrive at a definitive statement because that statement is actually based on emotion and beliefs, rather than based solely on scientific theories, methods, and measurements used by scien-

CHAPTER 17. TRANSLATORS

tists to test theories. Impairment is not a theory.

Then, after the numeric value of the observed concentration at the edge has been established and the 50% tic is located, fill in two more points. On the impairment side, think about the following story, familiar to many but that gives it impact. You are, with one hand offering someone a glass of water and the person knows the concentration of arsenic in that glass of water, and, with the other hand offering the person a thousand dollars to drink that glass of water, in front of everybody watching. When that person refuses the thousand dollars then you can conclude that the water in the glass is very impaired but not necessarily too dangerous to touch, only too dangerous to drink, and arbitrarily decide, for the purpose of establishing a translator, that the probability for this arsenic concentration is 100%. How much bad is tolerated, that's another way to think. Now on the other side, simply accept that zero concentration is 0%. Now reflect that this process of completing the graph is really that simple and easy. And, the process is based on information derived with science, more or less, but, completely relies on human emotion and all that emotion implies, when actually selecting numbers for a translator.

Finally, draw a line through the three points. You now have a simple translator. If that line is a straight line, it can be described with two numbers, the slope and the intercept. To force a straight line, draw through the origin and the 50% point only. The translator expressed as a slope and intercept are used in the computer code illustrations, found below.

Chapter 18

Two Databases

Two databases are needed now, one contains observational data for water bodies, the other water quality standards expressed as translators. The observations and the translators are associated with their respective water bodies, beneficial uses, method, et cetera.

```python
# this file name: C_18_assessment_data_doit.py
# dependencies:   C_18_assessment_data.f90
# to execute:
#           python C_18_assessment_data_doit.py
# This example code illustrates:
#           using Python to automate tasks
import os
os.system( \
    'gfortran -o D_18_run_c.out C_18_a_data.f90')
os.system('./D_18_run_c.out')
```

```fortran
! this file name:   C_18_a_data.f90
! dependencies: none
! to compile:
!       gfortran -o D_18_run.out
!                 C_18_a_data.f90
! to execute:   ./D_18_run.out
! This example code illustrates:
```

CHAPTER 18. TWO DATABASES

```
!          generate assessment data
      program C_18_assessment_data

      character*256 fa, fb
      character*200 crec
      character cha

      fa = 'E_18_translators.txt'
      fb = 'E_18_observations.txt'

   numtrans = 5000
! TRANSLATORS
      open(11, file=fa)
      iseed = 1257
      iran = IRAND(iseed)
      do j = 1, numtrans
         do k = 1, 200
            n = 1 + MOD(IRAND(),26)
            crec(k:k) = char(n + 64)
         end do
         do k = 1,10
            m = (k-1)*20 + 1
            crec(m:m) = char(32)
         end do

   crec(101:120) = '                    '  ! empty
   crec(121:140) = '         1.0        '  ! slope
   crec(141:160) = '        -1.0        '
                                           ! intercept

         write(11,'(a)') crec
      end do
      close(11)

      maxobs = 100
! OBSERVATIONS
      open(11, file=fa)
      open(12, file=fb)
      do j = 1, numtrans
         read(11,'(a)') crec

! mess up the data with
```

```
!    "no translator for some observations" and
!    "no observations for a translator"
         k = MOD(IRAND(),20)
         if( k .LT. 3 )then
!           no translator
               icha = ICHAR(crec(5:5))
               if(icha .GT. 80)then
                  crec(5:5) = CHAR(icha - 1)
               else
                  crec(5:5) = CHAR(icha + 1)
               end if
         end if

!        skip if no observations for this translator
         if( k .LT. 17 )then

            nobs = 3 + MOD(IRAND(),maxobs)
            do k = 1, nobs

            id = 1 + MOD(IRAND(),100000) + 400000
            re = 1.0 + 0.05 * REAL(MOD(IRAND(),20))
            ri = 0.5 + 0.01 * REAL(MOD(IRAND(),20))

      write(crec(101:120),'(i20)')id      ! date-time
      write(crec(121:140),'(f20.5)')re    ! value
      write(crec(141:160),'(f20.5)')ri    ! represen-
                                          !   tativeness

               write(12,'(a)')crec
            end do

         end if

      end do

      close(11)
      close(12)

      end
```

This and previous chapters have introduced and illustrated most of the manipulative tools that will be used here in the computer code that illustrates assessment using the concrete example of water quality. These manipulative tools are general in scope, and are well described in text books and the Internet. Now all this is assembled together, to produce assessment results from observational data and standards. This is done in the code in the next chapter. No narrative description is included, the reader by now has seen all the tools, and all the code illustrating each tool, so the reader will easily recognize what is happening when reading the code itself.

Chapter 19

The Zipper

This is what you may point to when you're asked, how does this case study end. You may point to a computer program that illustrates automating the assessment part of water quality assessment, using all the parts and ideas discussed above. This case study, however, is about thinking. When asked, point to the thinking processes gone through to arrive here. This case study is a journey, arriving at, perhaps, some new ways for an individual to think about assessment, quality, truth, and the use of computer technology to augment human thinking. And, for those who feel the frustration of exclusive privileged access for acquiring skills and knowledge dealing with computer technology, point to the start of a road to computer literacy, and point to yourself, already now several steps along that road.

Summarizing what follows, two databases are generated, representing a large amount of water quality monitoring data, and, water quality assessment standards and criteria expressed as a translator. These two databases are processed in a way so that a particular datum is aligned with the corresponding translator, Bayes Theorem is applied, corrections for imperfect data applied, and out comes the best possible value of the current likelihood of impairment for the particular water body, designated use, method, and based on all historic data including the latest datum.

```
! this file name:   C_19_illustrate.f90
! dependencies: none
! to compile:   gfortran -o D_19_run.out
!                      C_19_illustrate.f90
! to execute:   ./D_19_run.out
```

CHAPTER 19. THE ZIPPER

```
! This example code illustrates:
!    generate simulated data and process data
      program illustrate
      character*256 fa,fb,fc

      fa = 'E_19_simplisticTRAN.txt'
      fb = 'E_19_simplisticOBSR.txt'

      call makedata(fa,fb)

      fc = 'E_19_simpleRESULTS.txt'

      call processdata(fa,fb,fc)

      end
!_____

      subroutine makedata(fa,fb)

      character*256 fa,fb
      character*200 crec
      character cha

      numtrans = 5000
! TRANSLATORS
      open(11,file=fa)
      iseed = 1257
      iran = IRAND(iseed)
      do j = 1, numtrans
         do k = 1, 200
            n = 1 + MOD(IRAND(),26)
            crec(k:k) = char(n + 64)
         end do
         do k = 1,10
            m = (k-1)*20 + 1
            crec(m:m) = char(32)
         end do

      crec(101:120) = '                    '  ! empty
      crec(121:140) = '         1.0        '  ! slope
      crec(141:160) = '        -1.0        '
                                              ! intercept
```

```
            write(11,'(a)') crec
         end do
         close(11)

         maxobs = 100
! OBSERVATIONS
         open(11, file=fa)
         open(12, file=fb)
         do j = 1, numtrans
            read(11,'(a)') crec

! mess up the data with
! "no translator for some observations" and
! "no observations for a translator"
            k = MOD(IRAND(),20)
            if( k .LT. 3 )then
!              no translator
               icha = ICHAR( crec(5:5))
               if(icha .GT. 80)then
                  crec(5:5) = CHAR(icha - 1)
               else
                  crec(5:5) = CHAR(icha + 1)
               end if
            end if

!        skip if no observations for this translator
            if( k .LT. 17 )then

               nobs = 3 + MOD(IRAND(),maxobs)
               do k = 1, nobs

                  id = 1 + MOD(IRAND(),100000) + 400000
                  re = 1.0 + 0.05 * REAL(MOD(IRAND(),20))
                  ri = 0.5 + 0.01 * REAL(MOD(IRAND(),20))

         write(crec(101:120),'(i20)') id      ! date-time
         write(crec(121:140),'(f20.5)') re    !   value
         write(crec(141:160),'(f20.5)') ri
                              ! represersentativeness

                  write(12,'(a)') crec
```

CHAPTER 19. THE ZIPPER

```
            end do

         end if

     end do

     close(11)
     close(12)

     end subroutine
```
!──
```
     subroutine processdata(fa,fb,fi)
     character*256 fa,fb,fc
     character*256 fd,fe,ff,fg,fi

     fd = 'E_19_TRAsort.txt'
     fe = 'E_19_TRAuniq.txt'
     ff = 'E_19_OBSsort.txt'
     fg = 'E_19_OBSuniq.txt'

     call system('sort ' // fa // ' > ' // fd)
     call system('uniq ' // fd // ' > ' // fe)
     call system('sort ' // fb // ' > ' // ff)
     call system('uniq ' // ff // ' > ' // fg)

     call zipper(fe,fg,fi)

     end subroutine
```
!──
```
     subroutine zipper(fe,fg,fi)

     character*256 fe,fg,fi
     character cuf

     INQUIRE(FILE=fe,SIZE=lenfileT)
     INQUIRE(FILE=fg,SIZE=lenfileO)

     nin = 11
     open(nin,file=fe)
     cuf = char(32)
```

```
        lenrec = 0
        do while (ichar(cuf) .NE. 10)
           call fgetc(11,cuf,istat)
           if(istat.NE.0)go to 999
           lenrec = lenrec + 1
        end do
        close(nin)

        numrecT = lenfileT / lenrec
        numrecO = lenfileO / lenrec

        call zipit(lenrec,numrecT,numrecO,fe,fg,fi)

999     continue

        end subroutine
```
!_____

```
        subroutine zipit(lenrec,numrecT,  &
                         numrecO,fe,fg,fi)

        character*256 fe, fg, fi

        character(lenrec) rra, rrb
        character(lenrec), allocatable :: ra(:)
        character(lenrec), allocatable :: rb(:)
        character*80 ctemp, dtemp

        allocate(ra(numrecT))
        allocate(rb(numrecO))

        nin = 11
        open(nin, file=fe)
        do j = 1, numrecT
           read(nin,'(a)')ra(j)
        end do
        close(nin)

        open(nin, file=fg)
        do j = 1, numrecO
           read(nin,'(a)')rb(j)
```

CHAPTER 19. THE ZIPPER

```
      end do
      close(nin)
```

! position	1:20	21:40	41:60	61:80
! OBS	jurisdiction	where	use	method
! TRA	jurisdiction	where	use	method
! length	20	20	20	20
! index	1	2	3	4

! position	81:100	101:120	121:140
! OBS	when	date/time	value
! TRA	when	date/time	slope
! length	20	20	20
! index	5	6	7

! position	141:160	161:180
! OBS	representativeness	QA
! TRA	intercept	QA
! length	20	20
! index	8	9

! position	181:200	201	
! OBS	reference	(end)	
! TRA	reference	(end)	
! length	20	1	(total = 201)
! index	9	10	

```
      nou = 12
      open(nou, file=fi)

      ja = 1      ! record-number for TRANSLATOR
      jb = 1      ! record-number for OBSERVATION
      kount = 0!  number of OBSERVATIONS processed
                  ! in one batch
      ini = 0     ! flag for processing within
                  ! a batch
      ido = 0     ! flag to signal UPDATE and WRITE
                  ! SUMMARY in a batch
```

CHAPTER 19. THE ZIPPER 132

```
! Top of loop over all data.
! Processed as a "merge" to bring together
! matching TRANSLATORS and OBSERVATIONS.
      do while( (ja <= numrecT) .AND.  &
                                  (jb <= numrecO) )

         write(nou,*)
      write(nou,*)'#### NEXT ####   ini,kount= ', &
                                            ini,kount
         ctemp = ' '
                ! character array to hold output
         dtemp = ' '

         if(jb .LT. numrecO)then
           write(ctemp( 5: 8),'(i4)')ja
           ctemp(10:17) = ra(ja)(1:8)
           write(ctemp(20:23),'(i4)')jb
           ctemp(25:32) = rb(jb)(1:8)
           write(ctemp(35:38),'(i4)')jb+1
           ctemp(40:47) = rb(jb+1)(1:8)
         else
           write(ctemp( 5: 8),'(i4)')ja
           ctemp(10:17) = ra(ja)(1:8)
           write(ctemp(20:23),'(i4)')jb
           ctemp(25:32) = rb(jb)(1:8)
         end if
         dtemp(22:69) = ctemp(1:47)
         write(nou,*)dtemp(1:70)

         kap = 0 ! record index step for
                 !   TRANSLATOR in array ra(ja)
         kbp = 0 !   OBSERVATION in array rb(jb)

      if( ra(ja)(1:100) .LT. rb(jb)(1:100) )then
write(nou,*)  &
' TRA is .LT. OBS <<<<<<<<   trans has no obs'
         kap = 1
         ini = 0
         ido = 0
      end if
```

CHAPTER 19. THE ZIPPER

```
               if( (jb .LT. numrecO) .AND. &
                   (ra(ja)(1:100) .EQ. rb(jb)(1:100) ) &
                   .AND. (rb(jb)(1:100) .EQ. &
                   rb(jb+1)(1:100) ) )then
 write(nou,*)' TRA is .EQ. OBS   in middle ====='
                   kbp = 1

                   if(ini .EQ. 0)then
                     ini = 1
                     kount = 0
                     postrep = 0.5
                     postprob = 0.5
                     read(ra(ja)(121:140),'(f20.5)')slope
                   read(ra(ja)(141:160),'(f30.5)')tercept
                   end if

                   ido = 1
                     !     UPDATE

               end if

               if( (jb .LT. numrecO) .AND. &
                   (ra(ja)(1:100) .EQ. rb(jb)(1:100) ) &
                   .AND. (rb(jb)(1:100) .NE. &
                   rb(jb+1)(1:100) ) )then
 write(nou,*)' TRA is .EQ. OBS   at end group ==='
                   kbp = 1

                   ido = 2
                     !   UPDATE and WRITE SUMMARY

               end if

               if( (jb .EQ. numrecO) .AND. &
                   (ra(ja)(1:100) .EQ. rb(jb)(1:100) ) )then
 write(nou,*)' TRA is .EQ. OBS at end array ===='
                   kap = 1
                   kbp = 1

                   ido = 2
                     !   UPDATE and WRITE SUMMARY
```

CHAPTER 19. THE ZIPPER

```
            end if

            if( ra(ja)(1:100) .GT. rb(jb)(1:100))then
write(nou,*) &
' TRA is .GT. OBS  >>>>>>>>>>> obs has no trans'
                kbp = 1
                ini = 0
                ido = 0

            end if

            if(ido .GT. 0)then
            ! UPDATE
                write(nou,*)'        UPDATE'
                kount = kount + 1

                read(rb(jb)(121:140),'(f20.5)')zval
                read(rb(jb)(141:160),'(f30.5)')zrep
                zprb = slope * zval + tercept

! tame the input
                if(zprb.gt.0.95)zprb = 0.95
                if(zprb.lt.0.05)zprb = 0.05
                if(zrep.gt.0.95)zrep = 0.95
                if(zrep.lt.0.05)zrep = 0.05

                priorprob = postprob
                priorrep  = postrep

! Bayes Rule update, probability
                prob = zprb
                xprior = 1.0 - priorprob
                xprob  = 1.0 - prob

                pnum = prob * priorprob
                pden = pnum + xprob * xprior

                postprob = pnum / pden

! modify update probability
! with representativeness
                rep = zrep
```

CHAPTER 19. THE ZIPPER

```
            if(rep.gt.0.99)rep = 0.99
            if(rep.lt.0.01)rep = 0.01
            postprob = rep * postprob + (1.0 - rep) &
                                       * priorprob

! modify update probability for responsiveness,
! tame the update
            if(postprob.gt.0.95)postprob = 0.95
            if(postprob.lt.0.05)postprob = 0.05

! bayes rule update, representativeness
            rep = zrep
            xprior = 1.0 - priorrep
            xprob = 1.0 - rep

            pnum = rep * priorrep
            pden = pnum + xprob * xprior

            postrep = pnum / pden

! tame the responsiveness
            if(postrep.gt.0.95)postrep = 0.95
            if(postrep.lt.0.05)postrep = 0.05

! write it out
        write(nou,*)'zprb,zrp,postprob,postrep= ', &
                     zprb,zrep,postprob,postrep

            if(ido .EQ. 2)then
              ! WRITE SUMMARY
   write(nou,*)'      WRITE SUMMARY for ',ra(ja)(1:8)

   write(nou,*)'SUMMARY kount,postprob,postrep=', &
                        kount,postprob,postrep
            write(nou,*)

                kount = 0
                ini = 0
                ido = 0
            end if

          end if
```

```
                    ja = ja + kap
                    jb = jb + kbp

            end do
! End of all TRANSLATORS and OBSERVATIONS that
! have a match.

! Process the left-overs.
        do while( ja <= numrecT )
            write(nou,*)'#### NEXT ####'
            write(nou,*) &
  ' TRA is ALONE       <<<<<<<<     trans has no obs'
            ctemp = ' '
            dtemp = ' '
            write(ctemp( 5: 8),'(i4)')ja
            ctemp(10:17) = ra(ja)(1:8)
!           write(ctemp(20:23),'(i4)')jb
!           ctemp(25:32) = rb(jb)(1:8)
            dtemp(22:69) = ctemp(1:47)
            write(nou,*)dtemp(1:70)
            ja = ja + 1
        end do

        do while( jb <= numrecO )
            write(nou,*)'#### NEXT ####'
            write(nou,*) &
  ' OBS is ALONE       >>>>>>>>>>> obs has no trans'
            ctemp = ' '
            dtemp = ' '
!           write(ctemp( 5: 8),'(i4)')ja
!           ctemp(10:17) = ra(ja)(1:8)
            write(ctemp(20:23),'(i4)')jb
            ctemp(25:32) = rb(jb)(1:8)
            dtemp(22:69) = ctemp(1:47)
            write(nou,*)dtemp(1:70)
            jb = jb + 1

! Bottom of loop over all data.
        end do

    write(nou,*)'Finished, ja,numrecT,jb,numrecO=', &
```

CHAPTER 19. THE ZIPPER

```
                         ja,numrecT,jb,numrecO
    close(nou)

    deallocate(ra)
    deallocate(rb)

    end subroutine
!
```

The output of the above illustration in this case study serves only one purpose, to show that all this works. All the ideas are gathered together, after testing the parts of the computer code, allowing us to grasp now and then some understanding by going through the process of taking an idea, thinking how the idea could be expressed as computer code which illustrates the idea. Examining the output confirms our ideas are on the right track, or, demonstrates something is not working, either the idea, or our understanding of the idea, or our articulating our understanding by writing computer code to test the idea. Once everything works, our understanding has grown, and, perhaps we are more comfortable with the third greatest event for human thinking.

It's fine to say this zipper works for simulated data, but, that data came from a random number generator, and, no evidence yet that all this works in the real world, with real water bodies, real causes of impairment, and real translators. Good criticism. Behind this criticism is the feeling that even though a good faith effort was made to understand Bayes Theorem, sorted data, and other details, the feeling of belief is lacking. All this is still too abstract and Bayes Theorem is so abstractly complicated that belief is difficult to achieve.

Real data for real water bodies, along with the existing water quality standards for the real water bodies, and along with historic assessment decisions for these real water bodies, are available. So, pick a few to run through the zipper to see how real data works. Alter the code by dividing the zipper into two parts, the first part generates much smaller data files, and, the second runs the zipper itself. With a text editor, insert real data and insert real translators for that real data. Then put the data files with real data added through the zipper, and examine the resulting output file. Change real data and real translators and run again, and examine what happens in the resulting output. By focusing on real data for real water bodies and real water quality standards converted into real translators, and not focusing on the intricacies of the approach, a feeling develops about believing all this works or not. This evaluation phase, with real data, shows dramatically how human thinking, and human beliefs and feelings, can be augmented, assisted, enhanced, by the interaction of humans and computers. This evaluation phase is definitely within the realm of the third great evolution of human thinking, people using computer technology to assist in thinking. Furthermore, we know human emotions can be influenced by talking, listening, writing, and reading. This case study, by bringing into the assessment process all things emotional, is in step with the historic evolution of verbal and written communication and thinking, which also include emotions and beliefs.

The remainder of this chapter is a tangent more appropriately included here rather than earlier in the journey. It is more ideas about computer languages, because the computer language is how the human mind interacts intimately with computers. The examples of computer code included in this book only touch lightly on what computer languages are capable of doing, which is appropriate for the reader who has little or no experience in coding.

The code, above, is written in a procedural computer language. People learning about computer technology and applying the technology by using computer languages, frequently have questions about the pros and cons of different computer languages. They also have questions about the pros and cons of particular computers, such as brand, model, capabilities.

A procedural language was chosen for this illustration because a procedural language explicitly goes through computer code one instruction at a time, and it is usually obvious what is happening to the values contained in variables mentioned in each instruction. This makes procedural languages, in general, easy to follow when reading the code, and by assumption easy to understand. As a result, again in general, a procedural language is easy to learn and use, because everything happens one small step at a time.

When using computers to assist, or augment, our thinking, a procedural computer language is a natural place to start. There is a limited amount that the human mind can grasp at one time, genius obvious has a bigger grasp than most of us. The human mind can take a concept, mull about it, and if need be, break it down into parts and steps, and if the parts and steps are simple enough to comprehend, then putting them together allows the human mind to understand the concept with only understanding of the parts and steps and how these are put together. Computer technology allows the human mind to observe how the concept behaves, by imitating the behavior with a computer. So a concept that is too much for a human mind to understand, such as gravity, becomes understandable in terms of the behavior of the concept. Using the gravity example, we ask why gravity and get no satisfactory answer. We ask how and get

CHAPTER 19. THE ZIPPER

satisfaction that even though we don't know or understand gravity, we understand enough to predict the flight of a thrown stone and, design, construct, and successfully send people to the moon and back.

As the years go by, computer technology and computer applications grow increasingly complex, both in what they do and how much information they handle. A time in history occurred when some people who think about computers realized that using only procedural languages easily results in inefficiencies and ineffectiveness, especially when managing large complex computer applications. So they mulled over this problem and came up with object-oriented language, to address their particular problem. This is an expansion and a specialization of computer languages, and, over time a host of similar situations arose, and, a specialized language was developed to address each of the problems. Two examples are HTML, a language specialized to manage Internet web pages, and, the historic example is C++, which extended the procedural language C to emphasize and facilitate object-oriented code.

When people consider the pros and cons of different computer languages, especially when these people desire to gain some skills and knowledge that would allow them to successfully write computer code, the widely used languages come to mind. Often today, object-oriented comes to mind first. So, an illustration of object-oriented code is given below. This illustration does, in miniature, what the above program does with water quality assessment, except, exceedingly simplified so the object-oriented approach may be clearly seen.

```
#   name of this file:   C_19_doitoo.py
#   to execute:  python C_19_doitoo.py
#   this program illustrates:
#     object-oriented coding that
#     runs a miniature of this case study

# needed to get time-date stamp
import os

# the following three functions cause external
# computer code to execute, and
# will be called from within this example
# of object-oriented code
def set_up_obs_data():
    os.system('gfortran C_19_BAZdata.f90')
```

CHAPTER 19. THE ZIPPER

```python
        os.system('./a.out')

def set_up_tra_data():
    os.system('gfortran C_19_BAZtrans.f90')
    os.system('./a.out')

def the_zipper():
    os.system('gfortran C_19_BAZipper.f90')
    os.system('./a.out')

# start this illustration of object-oriented code,
# establish an object, name it DoBAZ for
#     Baysian Assesment Zipper
class DoBAZ():

# the following is required, it initializes
# an instance of this object
    def __init__(self):
            return None

# establish three methods within this object
    def dodata(self):
# this method calls a
#function established elsewhere
            set_up_obs_data()

    def dotran(self):
            set_up_tra_data()

    def dozipper(self):
            the_zipper()
# end of establishing this object, which
# contains three methods

# begin sequence of instructions,
# the first instruction is: initialize an instance
#      of the object, DoBAZ, and name it DB
DB = DoBAZ()
# the second instruction causes the object method
# to run, in this case, generate a file with
# observation data
DB.dodata()
```

CHAPTER 19. THE ZIPPER

```python
# generate a file with translator data
DB.dotran()
# run the zipper, write output to a file
DB.dozipper()
# put a time-date stamp
# at the end of the output file
os.system("date >> E_19_BAZipper_out.txt")
# end of illustration of object-oriented code
```

```fortran
! name of this file: C_19_BAZ_data.f90
! compile and execute:
!          automatically from C_19_doitoo.py
! this program illustrates: this case study,
!     create a miniature observation data file,
      program C_19_BAZ_data

      open(7, file='E_19_BAZdata.txt')
      do j = 1, 8
         write(7,*) j
      end do
      close(7)

      end
```

```fortran
! name of this file: C_19_BAZ_trans.f90
! compile and execute:
!          automatically from C_19_doitoo.py
! this program illustrates: this case study,
!     create a miniature translator data file,
      program C_19_BAZ_trans

      open(7, file='E_19_BAZtrans.txt')
      do j = 1, 8
         k = 9 - j
         write(7,*) k
      end do
      close(7)
```

end

```
! name of this file:   C_19_BAZipper.f90
! compile and execute:
!          automatically from C_19_doitoo.py
! this program illustrates:   this case study,
!    a miniature zipper,
  program C_19_BAZipper

     open(7, file='E_19_BAZdata.txt')
     open(8, file='E_19_BAZtrans.txt')
     open(9, file='E_19_BAZipper_out.txt')
     do j = 1, 8
        read(7,*) kd
        read(8,*) kt
        kz = kd + kt
        write(9,*) kz
     end do
     close(7)
     close(8)
     close(9)

  end
```

The essence of this miniature illustration is the following: object-oriented code is used to construct and manage a structure in our imagination, so we see the big picture, and can effectively work productively with this mental structure, usually within a large cooperative effort involving supervisors, teams, and individual coders. The mental structure is constructed with computer code. Using computer lingo, features of the code create objects and causes these objects to do things. An example, with a digital camera, take a picture. Then with software manipulate that picture. Likely the software you may find and use is written in the object-oriented approach, but not necessarily. The picture is an object described by the computer code. An example of an object is the contents of a JPEG file. But an object-oriented object is usually more than just data, it also contains methods, which manipulates the data part of the object.

CHAPTER 19. THE ZIPPER

With photo manipulating software, you can change just about anything in this digital photograph, such as color. The photo-software, if it follows the object-oriented approach, will have methods that do the actual work of changing what you want changed. In general, these methods do things that are complicated to the casual observer, and complicated even to the coder who is using object-oriented approaches to writing some code. So, this complication is packaged and hidden from the coder, unless of course the coder really wants to see it. The coder is then free to focus on the overall management of photos, what can be done to photos, and making all that happen, simply, and without worrying about details such as how to change color in a JPEG file.

Continuing this tangent, also appropriate here instead of earlier, is simultaneously coding in two or more computer languages when working on a large or complex project, which may require the special capabilities of more than one computer language.

In many of the listings, above, more than one computer language was used in an illustration. Usually it was Python causing other programs to execute. However, each program was executed separately and data transfer from one program to the next was by means of files. It may be desirable for two computer languages to share data while technically both are executing together, that is, started with one command to execute and then jumping back and forth between two computer languages. This is called interfacing between two computer languages. This is accomplished by coding such that one language is used to establish the place within the code where everything starts, and then continue where subroutines are coded in either computer language. The crucial feature is that all the subroutines and the main program can share data with each other seamlessly while executing and without the necessity of writing the data out to a file and then reading it back in, as it was done in many of the above coding examples.

The following listings illustrate mixed language programming for C and Fortran together.

```
#   This file name:     C_19_mixedSH.sh
#
#  dependencies:   part of a suite,
#           C_19_mixedSH.sh,
#           C_19_mixedC.c,
#           C_19_mixedF.f90
#  to compile:   execute this shell-script
```

```
# to execute:  see notes below
# This example code illustrates:
#       mixed language programming
#
#
# If the user writes and saves this file,
# then before it can be used, the following
# is needed to give this file permission
# to execute:
# in Terminal,    chmod u+x C_19_mixedSH.sh
#
# To execute this shell-script:  in Terminal,
#                         ./C_19_mixedSH.sh
#
# This shell-script compiles and creates an
# executable file, D_19_mix.out
#
# To execute the result, in Terminal,
#                              ./D_19_mix.out
#

gfortran -c C_19_mixedF.f90
gcc -c C_19_mixedC.c
gfortran -o D_19_mix.out C_19_mixedF.o \
                        C_19_mixedC.o
rm *.o # not necessary, just keeps
       # the directory cleaner
```

```
! this file name:  C_19_mixedF.f90
! dependencies:  part of a suite,
!        C_19_mixedSH.sh,
!        C_19_mixedC.c,
!        C_19_mixedF.f90
! to compile:  use C_19_mixedSH.sh
! to execute:  see notes in C_19_mixedSH.sh
! This example code illustrates:
!       mixed language program interfacing

      program C_19_mixedF
```

CHAPTER 19. THE ZIPPER

```fortran
      integer ii, jj, kk
      common/ijk/ ii, jj, kk

      real*4 ff, gg, hh
      common/fgh/ff,gg,hh

      ii = 2
      jj = 3
      kk = 4
      ff = 0.5
      gg = 0.6
      hh = 0.7

      call abc()   ! call the C-function

      write(*,*)'ii,jj,kk =',ii,jj,kk
      write(*,*)'ff,gg,hh =',ff,gg,hh


      end
```
!————————————————————————————
```fortran
      subroutine one(nn) ! subroutine called by C

      integer ii, jj, kk
      common/ijk/ ii, jj, kk

      write(*,*)'ii,jj,kk,nn =',ii,jj,kk,nn

      ii = 1234
      jj = 2345
      kk = 3456
      nn = 4567

      return
      end
```
!————————————————————————————

CHAPTER 19. THE ZIPPER

```c
// this file name:    C_19_mixedC.c
// dependencies:   part of a suite,
//          C_19_mixedSH.sh,
//          C_19_mixedC.c,
//          C_19_mixedF.f90
// to compile:   use C_19_mixedSH.sh
// to execute:   see notes in C_19_mixedSH.sh
// This example code illustrates:
//       mixed language program interfacing

// Declare an external subroutine, that is,
// interface with Fortran, and,
// underscore in name is required.
extern void one_(int *nn);

// declare an external structure
extern struct
{
   int ii, jj, kk;
} ijk_;            // the underscore is required
                   // for interface

// declare an external structure
extern struct
{
   float ff, gg, hh;
} fgh_;

// provide a C-function called from Fortran
void abc_(void)    // the underscore required
{
   ijk_.ii = 2222;   // the underscore required
   ijk_.jj = 3333;
   ijk_.kk = 4444;

   int na = 5555;

   one_(&na);   // calls the external subroutine

   float x = (float)na;
   fgh_.ff = x;
   fgh_.gg = x;
```

```
    fgh_.hh = x;

    return;
}
```

A disclaimer is obligatory now. Mixed language programming is not usually covered in tutorials or textbooks, so we're on the edge of what is easy to find, easy to figure out, easy to do the first or second time. The above illustration works for the Linux system favored here, that is, Crunchbang 11, and the gcc compiler that comes already installed, and the gfortran compiler that may be easily installed from the Internet using the apt-get tools that come already installed. No guarantee that anything different will work, let alone work perfectly first time as all the illustrations in this book.

Chapter 20

Watersheds and Finding a Segment

Sometimes, say when some people are actually going to implement an automated water quality assessment software package, based on this case study, a whammy is thrown into the works, usually to cause disruption such as hoping the whammy can not be addressed so the whole project dies. In this specific case, the whammy is: I want to collect and submit data that comes from anywhere in a watershed or shoreline, and I know classified segments are only a small fraction of a watershed. So include in the automation the ability to pick an arbitrary location in the watershed and assign to that location the applicable standards. If you can't do this, I'm out.

The following program, makedag, creates a representation of watersheds and shorelines as a branching tree. Picking any point in that branching tree, if no standards are assigned to that point, then go with the flow to find the first point downstream where standards have been assigned, and use those standards for assessing the water quality for the point.

The following code is an ad-hoc solution to this issue. The name appearing, DAG, comes from Directed Acyclic Graph, which is a formal approach to a subject well studied and documented. This ad-hoc solution handles all types of branching tree structures, including points where more than two twigs meet, points where a twig diverges into two or more down-stream twigs, often called braided structures. So this illustration shows that some times ad-hock computer code can be useful. There are situations where just finding and figuring out a formal approach and whether it is capable of addressing your needs takes more time and energy than

CHAPTER 20. WATERSHEDS AND FINDING A SEGMENT

just coding in an ad-hock manner, however, an elegant formal approach will likely be infinitely better coding-wise than an ad-hoc approach which tends to be messy and hard to understand. Accept any approach that works for you, keep in the back of your mind that a formal approach likely started by exploring in an ad-hoc manner, historically using pencil and paper, and today much more productively with computers.

```fortran
! this file name:  C_20_tryfibo.f90
! dependencies: none
! to compile:   gfortran -o D_20_runfibo.out
!                         C_20_tryfibo.f90
! to execute:   ./D_20_runfibo.out
! This example code illustrates:
!           Fibonacci search
      program tryfibo

      character*256 fin , fsort

      fin = 'E_20_Ydata.txt'
      fsort = 'E_20_Ysort.txt'

      lenseg = 20
      numseg = 50

      call dofibo(lenseg,numseg,fin , fsort)

      end
```

```fortran
      subroutine dofibo(lenseg,numseg,fin , fsort)
      character*256 fin , fsort
      character(lenseg) carr , darr
      dimension carr(numseg)

! numseg is the number of elements in the list ,
! search the list for a given element

! create test data
      iseed = 987
      x = RAND(iseed)
```

```
            open(12,file=fin)
            do j = 1, numseg
               x = RAND()
               write(12,'(f20.10)')x
            end do
            close(12)

! sort test data
            call system &
            ('sort E_20_Ydata.txt > E_20_Ysort.txt')

! read sorted test data into memory
            open(12,file=fsort)
            do j = 1, numseg
               read(12,'(a)')carr(j)
               if(j.EQ.20)darr = carr(j)
            end do
            close(12)

! write out sorted test data
            do j = 1, numseg
               write(*,*)j,' ',carr(j)
            end do

! preliminary, initialize the Fibonacci search.
         nf2 = 1
         nf1 = 1
         nf = 2
! Find the Fibonacci Number that is larger than
! the number of records in the dictionary file.
         do while (nf.LT.numseg)
            nf2 = nf1
            nf1 = nf
            nf = nf1 + nf2
         end do
! save
         initfibo = nf
         initfibo1 = nf1
         initfibo2 = nf2

! preliminary finished, begin extensive test,
!          part 1, find each element
```

CHAPTER 20. WATERSHEDS AND FINDING A SEGMENT

```
!              part 2, find impossible element

! test each entry and test impossible entries
         do kk = 1, 2
! when kk == 1 then entry will be found
! when kk == 2 then entry will NOT be found,
!              that is, doesn't exist
         do jj = 1, numseg

! darr is the element to be found, taken from the
! list so it will be found;
! carr() is the list of elements, sorted
            darr = carr(jj)
            if(kk .EQ. 2)then
!      change a character in the middle to
!      make it impossible to find
               darr(16:16) = 'A'
            end if

!            initialize the Fibonacci search
            nf   = initfibo
            nf1  = initfibo1
            nf2  = initfibo2
            i    = 0
            noff = 0

! the actual Fibonacci search algorithm
            ifound = -1
            do while (nf.GT.1)
               i = MIN(noff+nf2, numseg)
                  write(*,*)'i,noff=',i,noff

! the comparisons here are ASCII order
               if(darr.LT.carr(i))then
                  nf = nf2
                  nf1 = nf1 - nf2
                  nf2 = nf - nf1
               else if(darr.GT.carr(i))then
                  nf = nf1
                  nf1 = nf2
                  nf2 = nf - nf1
                  noff = i
```

```
                    else
                        write(*,*)' FOUND IT ',i
                        ifound = i
                        go to 360
                    end if
                end do
! end of Fibonacci search
                write(*,*)'NOT FOUND ::::::::::::: ',jj
                ifound = 0

360             continue

                write(*,*)'found   kk,jj ,i= ',kk,jj ,i

            end do
        end do

        end subroutine
```

```
! this file name:  C_20_makedag4.f90
! dependencies: none
! to compile:  gfortran -o D_20_run.out
!                       C_20_makedag4.f90
! to execute:  ./D_20_run.out
! This example code illustrates:
!    create a branched tree
        program makedag4

        character*256 fdag, fflo, fab, fsim, &
                                fsort, flink

        fdag  = 'E_20_Ztrydag.txt'
        fflo  = 'E_20_Ztryflo.txt'
        fab   = 'E_20_Zabbrev.txt'
        fsim  = 'E_20_Zsimula.txt'
        fsort = 'E_20_Zsorted.txt'
        flink = 'E_20_Zlinked.txt'

        numseg = 50
        lenseg = 20
```

CHAPTER 20. WATERSHEDS AND FINDING A SEGMENT

```
        call makedata(numseg,fdag)
        call makeflow(numseg,fdag,fflo,fab)
        call simulate(numseg,lenseg,fab,fsim)
        call system &
   ('sort E_20_Zsimula.txt > E_20_Zsorted.txt')

!  the following and similar statements are
!  normally for debugging, however used here for
!  output that demonstrates this code works
!  rather than output that is pretty
        write(*,*)'at A'
        call linkdag(numseg,lenseg,fsort,flink)

        end
```

```
        subroutine linkdag(numseg,lenseg, &
                                 fsort,flink)
!  take the sorted from-to-DAG-simulated, and
!  "link it", that is, convert to a format where
!     the "from-column" contains the alphanumeric
!                representation of a segment
!     the "simulated-DAG" is sorted,
!     the "to-column" is replaced with the array
!                index of the segment that has
!                the same name

        character*(*) fsort,flink

        character(lenseg) carr, darr
        dimension carr(numseg+1)
        character chu
                    write(*,*)'at B'
        numsegP = numseg
        open(12,file=fsort)
        j = 1
 10     continue
        read(12,'(a)',end=30)darr(1:lenseg)
           carr(j) = darr
                    write(*,*)j,' ',carr(j)(1:lenseg)
```

CHAPTER 20. WATERSHEDS AND FINDING A SEGMENT 154

```
            j = j + 1
         go to 10
30       continue
         close(12)
                              write(*,*)'at C'

! loop through all segments in the second column
! of the DAG, find where the name matches with
! the first column, when a match is found then
! replace the name in the second column with the
! array index-number of the match in the first
! column.

! Initialize the Fibonacci search.
      nf2 = 1
      nf1 = 1
      nf = 2
! Find the Fibonacci Number that is larger than
! the number of records in the dictionary file.

         do while (nf.LT.numsegP)
            nf2 = nf1
            nf1 = nf
            nf = nf1 + nf2
         end do
! save
         initfibo = nf
         initfibo1 = nf1
         initfibo2 = nf2
                              write(*,*)'at D'
! loop over all segment-names in the second column
         lenhalf = lenseg / 2
         leng = lenhalf - 1
         do jj = 1, numsegP

      darr(1:leng) = carr(jj)(lenhalf+1:lenhalf+leng)
                        write(*,*)'   jj= ',jj
                        write(*,*)darr(1:leng)

!           initialize the Fibonacci search
            nf   = initfibo
            nf1  = initfibo1
```

CHAPTER 20. WATERSHEDS AND FINDING A SEGMENT 155

```
            nf2  = initfibo2
            i    = 0
            noff = 0

            ifound = -1
            do while (nf.GT.1)
               i = MIN(noff+nf2, numsegP)
               if (darr(1:leng) .LT.       &
              carr(jj)(lenhalf+1:lenhalf+leng)) then
                  nf = nf2
                  nf1 = nf1 - nf2
                  nf2 = nf - nf1
               else if (darr(1:leng) .GT.  &
              carr(jj)(lenhalf+1:lenhalf+leng)) then
                  nf = nf1
                  nf1 = nf2
                  nf2 = nf - nf1
                  noff = i
               else
                  ifound = i
                  write(carr(jj) &
                  (lenhalf+1:lenhalf+leng),'(i9)')i
                  go to 360
               end if
            end do
            write(*,*) 'NOT FOUND ::::::::::: ', jj
360         continue

         end do
                                     write(*,*)'at E'
! write out the linked-DAG
         open(14, file=flink)
         do j = 1, numsegP
            write(14,*)carr(j)
         end do
         close(14)
                                     write(*,*)'at F'

         end subroutine
```

!_____

```
         subroutine simulate(numseg, lenseg, fab, fsim)
```

```
        character*(*) fab,fsim
        character(lenseg), allocatable :: cseg(:)
        dimension iseg(numseg+1)

        allocate(cseg(numseg+1))

        iseg = 0
        cseg = '.'

        open(12,file=fab)
        j = 1
10      continue
        read(12,*,end=30)ia
          iseg(j) = ia
          j = j + 1
        go to 10
30      continue
        close(12)

        lenhalf = lenseg / 2
        iseed = 2341
        j = IRAND(iseed)
        open(13,file=fsim)
        do j = 1, numseg+1
          do i = 1, lenhalf - 1
            k = 65 + MOD(IRAND(),25)
            cseg(j)(i:i) = char(k)
            m = iseg(j)
            if(m .EQ. 0)then
              cseg(j)(i+lenhalf:i+lenhalf) = '-'
            else
              cseg(j)(i+lenhalf:i+lenhalf) = &
                                     cseg(m)(i:i)
            end if
          end do
          cseg(j)(lenhalf:lenhalf) = char(32)
          cseg(j)(lenseg:lenseg) = char(32)
          write(13,'(a)')cseg(j)(1:lenseg)
        end do
        close(13)
```

```
      end subroutine
```

```
      subroutine makeflow(numseg,fin,fou,fab)
      character*256 fin,fou,fab
      dimension iseg(numseg+1)
      dimension iqla(numseg+1)
      character*50 ch

      iseg = 0
      iqla = 0

      open(14,file=fin)
      open(15,file='E_20_Zabbrev.txt')
```

! import the DAG which has already been converted
! from strings to consecutive numbers, such as a
! dictionary index

```
      j = 1
10    continue
      read(14,*,end=30)ia,ib,iq,ch
      if(ia .GT. 0)then
         write(15,*)ib    !abbreviated
         iseg(j) = ib
         iqla(j) = iq
         j = j + 1
      end if
      go to 10
30    continue

      close(14)
      close(15)
```

! start at an arbitrary segment, flow down-stream
! until a classified segment is reached; then
! apply the classification details to the starting
! segment

```
      open(16,file=fou)

      do n = 5, numseg + 1
```

CHAPTER 20. WATERSHEDS AND FINDING A SEGMENT

```
                j = n
                write(16,*)'next    n=',n
110             continue
                k = iseg(j)
                write(16,*)j,k,iqla(j)
                j = k
                if(j .NE. 0)go to 110
            end do

            close(16)

            end subroutine
```

!―――――――――――――――――――――――――――――――

```
            subroutine makedata(numseg,fa)
```

! Create a DAG where each element is a "from―to"
! flow―direction, where flow starts at a leaf and
! ends at a root.
!
! Each element contains two IDs that describe
! "where" the segment is located.
!
! Add to the DAG a flag indicating that segment is
! "classified". Later let that flag point to the
! classification details, such as what, use,
! method, when (season), and translator values.
!
! So, during processing (in the zipper or
! preparing files for the zipper), given any
! segment, walking the "DAG" will lead to a
! classification that is down―stream and is
! classified, which applies to all upstream
! segments.

```
                character*256 fa

                qlass = 0.2     ! probability of classified
                qleaf = 0.5     ! probability of a leaf
                qroot = 0.1     ! probability of a root
                qtwig = 0.1     ! probability of a twig
```

CHAPTER 20. WATERSHEDS AND FINDING A SEGMENT

```
          iz = 0

       open(12,file=fa)
       iseed = 1257
       x = RAND(iseed)
       j = 1
       do while( j .LE. numseg)
```
! given a segment, need to determine:
!   is segment classified?
!   if not, go downstream until find a
!   classified segment
! so need a "from—to" DAG, then
!   when sorted can find "from" easily, and then
!   using from—to, walk to classified segment;
!   need to convert "from" names when sorted into
!   consecutive numbers = index of record

```
              iq = 0
              x = RAND()
              if(x .LT. qlass)iq = 1

              if(j.EQ.1) then
```
!   root
```
                  write(12,*)j,iz,iq,' root '
                  write(12,*)j+1,j,iq,' stem '
                  j = j + 1
              end if

              if(j.GT.5) then
                 x = RAND()
                 if(x.LT.qroot)then
```
!   a new root
```
                    write(12,*)j+1,iz,iq,' root '
                    j = j + 1
                 end if

                 x = RAND()
                 if(x.LT.qtwig)then
```
!   a branch twig
```
                    it = INT(RAND() * REAL(j −1)) + 1
                    write(12,*)j,it,iq,' twig '
```

```
                    j = j + 1
                end if

                x = RAND()
                if (x.LT.qleaf) then
!               leaf
                    write(12,*) iz,j,iq,' leaf '
                    it = INT(RAND() * REAL(j-1)) + 1
                    write(12,*) j+1,it,iq,' twig '
                    j = j + 1
                end if

!               stem
                write(12,*) j+1,j,iq,' stem '
                j = j + 1

            else

!               stem
                write(12,*) j+1,j,iq,' stem '
                j = j + 1

            end if

        end do

! The basic DAG is complete

        close(12)

        end subroutine
!_____
```

This would be particularly useful if the "where" is not a descriptor or name but rather is a location expressed as latitude and longitude. Other ways can easily link latitude and longitude with a place descriptor.

Chapter 21

Where is Standard Deviation?

Our objective has been to think about performing an assessment, and while thinking use computer technology for assistance. We noted that Bayes Theorem may be worded in the language of assessment, and we followed the details of an illustration that simulated an assessment process. Thinking about the process simplistically, we notice that our assessment result is a conditional probability, which is, the probability of impairment given the evidence. When the value of this probability approaches one, we feel more certain that impairment is true. Likewise when the value approaches zero we feel more certain that impairment is false. This is the output of our assessment process. Now we'll give it a name, preponderance. That is, the preponderance of the evidence. When the preponderance exceeds a value selected with the same process for establishing translators, we can say the water body is impaired. At the other end is the input, another conditional probability, which is, the probability of the evidence given impairment. If we have a time sequence of values for this input, we observed it is a simple matter using Bayes Theorem in an update process to get the output value. So, we have an input values and an output value, and an automated machine in the middle, called Bayes Theorem.

All we need is a time sequence of input observed values, and, we've pushed all the politics, emotions, financial implications, beliefs, obstruction, obfuscation, good will, good faith, misunderstandings, ignorance, and unlimited other possibilities, on a translator, which may be designed so that observable and measurable values are converted into the needed input conditional probability.

A workable and very simple translator for a single observed or measured quantity or attribute is a straight line on a graph, where one axis is the observed value and the other axis is the desired probability. Put in some numbers and if every body agrees, the job is complete and the assessment machine can roll out results when data is rolled in.

However, in the real world, things are not this simple. The various stakeholders may fight tooth and nail over the numbers that define the line of the translator graph. Here is where the heavy equipment are brought in, which may include but not limited to all the statistical tools and concepts that exist, and new tools and concepts as they are developed. So, here is where Standard Deviation, Mean, Confidence Interval, Margin of Error, Type I and Type II Errors, and everything else, will be applied.

The vast arsenal of statistical and other techniques are essential when one question, raised here, is addressed. The question is, how is a translator written, or in other words, what are the instructions for determining numerical values for the probability of the evidence B given impairment A, written symbolically as, P(B | A). Finding a numerical value for this probability has been the focus of study from the present day all the way back to the beginnings of statistics. That means it is not easy. Look at Bayes Theorem, it is Bayes's attempt at determining the conditional probability.

Examining the translator, the evidence B includes all the observable, measurable, qualitative, quantitative, and results in any other format, that convey something about things and events in the real world, which here is narrowed to water quality. The impairment A is defined and expressed in the water quality standards and criteria, in a manner that allows evidence B to be converted into a probability of impairment A.

The first task is to understand and then articulate the essence of each water quality standard and criteria. The purpose of standards in the first place, hopefully, is to provide a way to use measurable or observable data for the purpose of determining whether or not a particular water body meets the applicable standards for the applicable beneficial uses assigned to that particular water body. So in reality, standards are not the task performed before all other assessment tasks, rather, standards need to be constructed so that methods implementing the standards take observational data and convert them into the probability of the water body evidence given the water body is impaired without any additional interpretations and data manipulation, which means that methodology and standards need to be constructed at the same time, and with open conversations between all involved all the time.

In addition to the translator, there are two other quantities that influence the assessment decision. They are, the representativeness, described above, and, the decision trigger level of the probability of impairment, now called the preponderance.

CHAPTER 21. WHERE IS STANDARD DEVIATION?

The representativeness manages problems with sampling, variation in the water body, and anything else that is not a mistake nor a quality assurance issue nor something recognizable that may be corrected on the spot. Representativeness addresses problems that cause a result to be less representative of the true condition of the water body.

The preponderance, which has a decision threshold level for the probability of impairment, is the quantity used to determine the assessment decision, usually worded as, the water body classified segment is impaired for the particular standard and beneficial use under consideration.

Some may think that the representativeness and the preponderance are opportunities to affect the assessment decision. This is a false hope. The only impact the representativeness and the preponderance has on the decision is to speed it up or slow it down. The translator is the only place that impacts the assessment decision.

When representativeness consistently has a value close to one then the Bayes update moves quickly, and if the value is close to zero the Bayes update moves slowly. This compensates for less than perfect observational data but is not intended for errors that are recognizable and fixable.

Where the data consistently show that impairment is correct, and the preponderance threshold decision level is set to 0.51, the threshold will be crossed quicker than if the threshold decision level is set to 0.91. This accommodates skepticism about the assessment process, where the skeptic wants a decision threshold value closer to one than closer to one half.

Who should establish the numeric values for representativeness and preponderance and what values are best? The same people and the same process to establish the translator and at the same time because in many people's minds the three are dependent upon each other. When push comes to shove, the translator is the sole determent of the assessment decision while the other two control only how fast or how slow the Bayes update process proceeds.

Strategically, having the numeric values of all three, translator, representativeness, preponderance, established, results in a deterministic and prescriptive assessment process, which many people will feel uncomfortable with because there are no opportunities along the way to "fix" something that they believe is not right. So, allow beliefs and emotions and politics to have ever present ability to modify the assessment process. Rather than establishing unchanging numeric values for representativeness and preponderance, establish default values. Let the people who do the sampling, the analysis, the shepherding of the tasks and data, substitute a different value than the default value, and, when substituting add to the record the rational for the particular substitution. What happens now is that the people who care most about assessment outcomes will moni-

CHAPTER 21. WHERE IS STANDARD DEVIATION?

tor closely any substitutions for the default values, and, while evaluating and strategizing will gain in-depth understanding what the translator is doing, which may lead to a needed revision of the standards and methods on which the translator is based, bringing into the process human emotions, beliefs, feelings, and arrive at an understanding in their minds and hearts of how much bad is tolerated when bad happens.

Getting back to standard deviation, take as an example, what is the diameter of this basketball that I have in my hand? So, put a basketball in your hand and ask the question. Now, somehow, you have measured the diameter of your basketball. What is the standard deviation, the mean, the confidence interval, the margin of error, and the type I and type II errors? With one measurement we can not proceed, so, make more measurements. Now we have enough information to calculate the mean and the standard deviation, but, we don't have enough information to calculate the others. Both confidence intervals and margins of error need to specify a desired probability, for example, what is the confidence interval so that we have a 50% probability that the next measurement we perform will have a result within the confidence interval. Now that we have specified all we need to specify and calculated results above, use all this data and answer the real question, which is, are we happy or unhappy with the basketball. To answer the real question with the approach described in this book, we need a translator that converts all the data we've assembled into an indication of our happiness, and here that indication is the probability of happiness, or, the preponderance of happiness taking it one step further.

Now, our happiness may depend on the standard deviation of our measurements, may depend on a particular size of a confidence interval, or, may only depend of the diameter of the basketball. If our happiness depends only on the diameter, then the standard deviation is not really needed for our deliberations.

If we measure the diameter of the basketball using some published method developed by a professional organization whose purpose is to standardize measuring the diameter of basketballs, we may be happy only if our measurements conform to the criteria established in the method. Then, standard deviation may be very important. We may get happiness knowing the measurements followed completely the method and met the quality assurance criteria mandated by the method. Or, we may get happiness by looking at the value of the standard deviation. Water quality assessment is analogous, we may need more than one indicator before we can arrive at an assessment decision. If so, then put all the requirements for an assessment decision in clear unambiguous sight before starting the process of developing standards, developing methods, et cetera. If standard deviation is critically important for making the assessment decision

CHAPTER 21. WHERE IS STANDARD DEVIATION? 165

then standard deviation value should be part of the translator. If standard deviation is important for screening data before the assessment decision then standard deviation should appear in the method where data is either voided or continues along the assessment path.

The devil's advocate has been listening, and says at this point, stop going in all these circles, we're smarter than that. You don't need translators, that just confuses things. Taking arsenic for example, professionals can tell you how much arsenic is bad for you and that's backed up by science and can be proved, so, use the numbers based on science and get on with it. Many agree, so we pursue that, and find a number called LD50 for ingestion of arsenic by rats, and we find lots of other similar numbers, that tell us how poisonous arsenic is. Now, which number do we choose for our water quality standard, certainly not the LD50 for rats. Looking at all these numbers related to LD50 we see that they express a probability. LD50 is the dose for which 50% of the subjects die. So, what number do we find desirable, the amount in water for which 20% of the people go blind, or 10% get cancer within five years, or 2% of children under the age of five develop high blood sugar? Using science gets understanding of what can and does happen, but people who put numbers into water quality standards must decide how much bad is tolerated when bad happens, which is an emotional decision, not a scientific fact. So probability already is used to express what can and does happen, and the translator is a way to make probability generally and universally and understandably usable for assessment purposes.

Now stepping back to look at the big picture, to look back at this journey we've taken and see where it has led us, and mulling all the bits and pieces we've seen. We're seeing the axioms, theorems, and lemmas of probability, seeing how Bayes Theorem can take data that is acquired periodically and update an estimate of the preponderance of impairment with the most recent data, seeing how all this can be part of a scheme for water quality assessment where a translator is needed to convert observational data into probability, seeing how two files can contain observational data and standards information for each water body segment, each beneficial use, each method, each seasonal and other circumstances, seeing how these two files can be sorted and merged in a process that brings the particular data together with the particular translator and update a sequence of periodic samples to arrive at the current best estimate of preponderance of impairment. And in the middle of all that, clearly recognize that human emotions and politics control the process and determine all outcomes and can not be circumvented nor neutralized. Seeing that observational data is not perfect, and data imperfection is addressed explicitly using representativeness. Now we see all this, and we stare at a big picture, and we see that

CHAPTER 21. WHERE IS STANDARD DEVIATION? 166

in the software illustrations in this book, the Bayes update process in the final illustration code is quite small, unnervingly simple, and always exactly the same for every observational datum and every translator. Then it becomes apparent that almost all the computer software illustration code in this case study is not needed in a real world implementation that uses the approach illustrated in this book. So take out the Bayes update code and throw away the rest of the code, which is almost all of it. Tweak the water quality standards and tweak the sampling and analysis and reporting protocols, so that everything works with a translator and a representativeness value and a preponderance threshold that are built into the standards and methods with no need for further information manipulation, further interpretation of regulations and guidance, and, no special cases and no exceptions. With computer automation, or by hand if need be, apply the Bayes update at the interface between analysis and reporting. The laboratory has the data. The translator, the representativeness, and preponderance threshold are built into the standards and methods which the laboratory has access to, the prior value of the preponderance of impairment is already in the final database, the reporting software takes the current value from the laboratory or other reporting entity, takes the prior value from the database, does a Bayes update, and puts the new current and latest value into the database. This may be done automatically, in real time, and the current status of all water bodies is available. If archived properly during the process, which means automated in sampling and analysis software, a paper trail also is available and current, making verification a matter of reviewing the electronic copy of the paper trail rather than the humongous task of finding and putting together all the bits and pieces of the paper trail for a particular reported value.

So we've seen a concrete example, simulated but that's fine, of using computer technology to augment our thinking, that is, come up with something potentially useful that we would not have come up with if we didn't have computer technology. We find our thinking is limited, especially when it comes to thinking about numbers and algebra and millions of numbers and millions of algebra expressions doing things with those numbers. Most of us need computer technology to think about these things that don't fit in our minds.

Chapter 22
Do It

The following compiles and executes every code illustration in this book, all at once.

```
# this file name:  everything.py
# dependencies:    all .py  .c  .f90  files
#                  in this book
# to execute:  python everything.py
# This example code illustrates:
#            testing the computer code in
#            this book, except Chapter 13
#            which has interactive app
# NOTE: copy this file and all
#       the .py .c .f90 files in this book into
#       a fresh new directory, then execute
#       this program
import os
import time
now = time.time()
print("START EVERYTHING")
print("1")
os.system('python C_1_doit.py')
print("8")
os.system('python C_8_ascii_doit.py')
print("9")
```

```
os.system('python C_9_step_doit.py')
print("10")
os.system('python C_10_dice_doit.py')
print("11")
os.system('python C_11_compare_doit.py')
print("12")
os.system('python C_12_heap_doit.py')
print("13")
a1 = "gfortran -o D_13_runseefile.out"
a2 = "  C_13_SeeFileBytes.f90"
a3 = a1 + a2
os.system(a3)
print("test SeeFileBytes manually,")
print("            it is interactive")
print("14")
os.system('python C_14_merge_doit.py')
print("16a")
os.system('python C_16_probability_doit.py')
print("16b")
a1 = "gfortran -o D_16_runcoin.out"
a2 = "  C_16_coinflip.f90"
a3 = a1 + a2
os.system(a3)
print("16c")
a1 = "./D_16_runcoin.out >"
a2 = "  E_16_seecoinflip.txt"
a3 = a1 + a2
os.system(a3)
print("18")
os.system('python C_18_assessment_data_doit.py')
print("19a")
a1 = "gfortran -o D_19_run.out"
a2 = "  C_19_illustrate.f90"
a3 = a1 + a2
os.system(a3)
os.system('./D_19_run.out')
print("19b")
os.system('python C_19_doitoo.py')
print("19c")
os.system('chmod u+x C_19_mixedSH.sh')
os.system('./C_19_mixedSH.sh')
os.system('./D_19_mix.out')
```

CHAPTER 22. DO IT

```
print("20a")
a1 = "gfortran -o D_20_runfibo.out"
a2 = "  C_20_tryfibo.f90"
a3 = a1 + a2
os.system(a3)
os.system('./D_20_runfibo.out > E_20_seefibo.txt ')
print("20b")
a1 = "gfortran -o D_20_rundag.out"
a2 = "  C_20_makedag.f90"
a3 = a1 + a2
os.system(a3)
a1 = "./D_20_rundag.out >"
a2 = "  E_20_seedagresults.txt"
a3 = a1 + a2
os.system(a3)
print("22")
a1 = "gfortran -o D_22_runwhat.out"
a2 = "  C_22_whathappens.f90"
a3 = a1 + a2
os.system(a3)
a1 = "./D_22_runwhat.out >"
a2 = "  E_22_seewhathappens.txt"
a3 = a1 + a2
os.system(a3)
print("END")
later = time.time()
lapsed = later - now
print("time for everything (seconds) =")
print(lapsed)
```

Reviewing where we've ended regarding thinking about water quality assessment, some of us, those who diligently mulled through most of the code, now likely have a different attitude about concepts and processes of water quality assessment than previously held beliefs.

We've seen that an implementation of some ideas presented here could involve a Bayes Theorem update process inserted in the interface between laboratories and other entities that generate monitoring data and some final database repository and final summary report, making the final report automatic to generate in real time giving the status of water quality in the nation continuously, like some clock high up on a tower showing the time.

We've seen that this approach puts all the politics, economics, beliefs, financial implications, and other such social forces, on to the translator, that converts specific monitoring data into a probability of impairment or attainment depending on the wording used. The translator is a concept that is universal to comprehend and universal to implement. If someone understands one example of a translator in the real world, that person easily understands any other translator.

Take a break and go back to Chapter 10 and revisit the dice game that Stooge and Shark play. Look at the place in the code where a variable named prob is given a numeric value. Notice that in both of the two passes through the data, the value given to prob is exactly the same, that is, 0.6 when Stooge believes that Shark's game is honest, and 0.4 when Stooge believes otherwise. This is the translator, and the only difference between the two passes through the data is how Stooge looks at and understands the evidence. In the first pass, Stooge uses common sense and sees whether or not Stooge is winning in the long run. Eventually the value of the variable named post reaches 0.01 and stays there, which indicates that the game is dishonest, and, in the process Stooge has lost money, and, learns the hard way. In the second pass, Stooge again uses common sense and sees whether or not Stooge is winning what Stooge expects to win in the long run if the game is honest including Shark's incentive by offering an uneven payoff. This time the value of post reaches 0.01 and stays there much quicker, and Stooge quits while Stooge is winning, not learning a lesson the hard way. Stooge in both passes through the data follows the truth each time, both of Stooge's ideas about assessing the game are true and correct, and neither is more true than the other. It is simply a matter of Stooge's point of view, Stooge's beliefs and feelings and understandings about the world, all which is true in Stooge's mind and heart. Winning in the long run may be the only option to test if everything else is unknowable, and, in the long run Stooge does learn, albeit the hard way. Winning in the long run what may be expected to win given a particular understanding of the game is an alternative belief, feeling, understanding about the world and no less or more true in Stooge's mind and heart. When this second option is available, it allows Stooge to learn quicker and perhaps quit while still winning.

Now consider taking this into a different realm, funding the government. The United States used tariffs at first, and later included income and payroll taxes. Congress knows that taxes may be used not only to get the money needed to fund the government's benefits, services, and other costs for providing what government provides, but also shape the behavior of society by encouraging and discouraging citizen's behaviors, and which for an example among many examples, cause wealth to distribute

CHAPTER 22. DO IT

in a particular manner. The idea of Shark offering an uneven payout can be interpreted several ways, one is that Shark is acting for some outside observer as if Shark acknowledges that the dice are dishonest and feels guilty about that and offers uneven payout to mitigate some of the dishonesty. Another interpretation is, Shark is offering an incentive to cause Stooge to play the game longer giving Shark a better opportunity to win. Now substitute the word fairness for dishonesty, and substitute the word taxes for rolling dice. Where this leads is, finance at the national level has an impact on society, and, in general the various ideas and approaches to finance are all true and correct just as Stooge's two different ideas illustrated in the two passes are true and correct, and even two interpretations of Shark's motives are each true and correct. The reason why all the different ideas and approaches to taxation are all true and correct is that each idea is formed and assessed using beliefs, feelings, convictions of truth, and all the emotions an individual has as a result of living in and interacting with a society, rather than formed and based on logic or mathematics or the laws of nature or the voice of authority. Logic and the rest are valuable for thinking which can help and influence what a person believes and feels, but not an absolute and certain process that leads to the correct beliefs, feelings, convictions of truth.

There is another long held belief and feeling, that reading, writing, and arithmetic are sufficiently important skills for citizens to acquire, for the purpose of coalescing around the general direction that society evolves. Societal methods such as taxation and the thousands of laws that say what a citizen or entity can and cannot do and what must or must not be done, are better understood by the people allowing informed agreement or disagreement, when reading, writing, and arithmetic are the basis of our educational system for all our children. Note that these skills exemplify the essence of the first two huge advances in thinking for humanity. Now we have the third huge advance. If we still believe that there are basic thinking skills that are important enough for fostering and maintaining the public good, then computer literacy in the same fashion as talking and listening literacy, and reading and writing literacy, should receive the same level of societal support as reading, writing, and arithmetic.

Computer literate individuals can relatively easily explore and mull the impact of various taxation schemes and many other issues impacting themselves and society, and, arrive at understandings and conclusions that allow the people to quit a particular approach while still ahead, and not learn the hard way which is losing in the long run before realizing a particular step in societal evolution had taken a less desirable path. Of course, a desirable path is an opinion or belief held by an individual and not necessarily universal among all citizens. Mulling and thinking is one thing that economists do with financial issues and other professionals do in various other fields, and so do individuals and entities that have, perhaps, exclusive privileged access to the where with all that economists and other professionals work with. Literacy helps people understand the world they have, and literacy helps people get the world they want.

The first huge advance in thinking likely helped humanity to cease living like an animal and start living in a civilization that more or less provided food and shelter. The second huge advance in thinking likely helped humanity to expand the scope of civilization along a path that leads to today, where the third huge advance is exerting its influence. And, with computer literacy a lot of exclusive privileged access to thinking about concepts and issues important to individuals and society may be circumvented, not putting experts and professionals out of a job, but allowing more in the general population to think better, that is, augmented or extended thinking with the aid of computers, which is possible if computer literacy is widespread. Just as reading and writing skills are somewhere between beneficial and necessary for our present society, computer literacy is or soon will be somewhere between beneficial and necessary.

Now we're back to the primary theme, which is thinking, with the third greatest advance of human thinking. With computer technology, we can think about ideas more complicated or larger than we can wrap our minds around.

Similar to the Dick, Jane, and Spot story in the first chapter, the following example code is the story of your life, greatly simplified of course.

```
! this file name:   C_22_whathappens.f90
! dependencies: none
! to compile: gfortran -o D_22_runwhat.out
!                     C_22_whathappens.f90
! to execute: ./D_22_runwhat.out >
!                     E_22_seewhathappens.txt
! This example code illustrates:
!              accumulated wealth over a lifetime
      program whathappens

         wealth     = 0.0
         gain       = 0.07    ! gain on investments
         saved      = 0.25    ! part put into wealth
         numyears   = 120
         wages      = 100.0
```

CHAPTER 22. DO IT

```
      expense    = 1.0 - saved ! needed to live
      costliving = expense * wages

   do j = 1, numyears

      if(j .LT. 22) then
      end if

      if(j .GE. 22 .AND. j .LT. 66) then
         start    = wealth
         retire   = gain * start
         tax      = wages * 0.20 + retire * 0.10
         thisyear = wages + retire - &
                                costliving - tax
         wealth   = start + thisyear
      end if

      if(j .GE. 66) then
         start    = wealth
         retire   = gain * start
         tax      = retire * 0.20
         thisyear = retire - costliving - tax
         wealth   = start + thisyear
      end if

      write(*,*) j, wealth

   end do

   end
```

This very simple computer program simulates your finances for your entire life. No Bayes Theorem, no statistics, no algebraic equations from finance or economics, just lots of simple addition, subtraction, and multiplication done by a computer in an understandable way. This very simple program, with the experience gained in the rest of this book, is trivial to alter, easy to understand, and so you can change the numbers, change how things are calculated, add all sorts of details such as IRAs, Social Security, medical care costs, tax law details, auto loans, real estate, corporations,

CHAPTER 22. DO IT

and, this list of details sometimes seems to extend to infinity. And, you can start with this simple code and extend and modify it so that it represents your current thinking or current life situation. Looking at the output of the computer code will change your thinking. If you attempt to test your ideas about your life without a computer, your mind will quickly become overwhelmed when the complexity reaches your limit, this is where analytical thinking comes in handy for those who find analytical thinking easy. Algebraic expressions for things such as compound interest and annuities also describe what happens. But, writing a few lines of code may take less effort than finding and understanding and putting together all the analytical approaches into your financial life situation.

Examining the output, we see that the person represented by the code will run out of money at age 96, and, running out of money at old age is not a desirable situation. What is your life expectancy, and, your ability to pay for your cost of living, for your entire life.

So, now that you have the ability to circumvent privileged exclusive access to knowledge critical to your life, go ahead, use this simple code to create a more complete and more accurate representation of your financial life and see what likely may happen if the world behaves as represented by your code. What will you think now when pundits, politicians, talk-radio, commentators, your friends, news broadcasters, financial advisors, talk about taxes, IRAs, inflation, insurance, mortgages, health care, investments, income disparity, the middle class, and a host of other ways that money impacts your life in our society. With help from a cost-effective salvaged computer and some free software, all under your complete control and in the privacy of your kitchen table, you can wrap your mind around these and other equally important issues. It's all about thinking.

www.ingramcontent.com/pod-product-compliance
Lightning Source LLC
Chambersburg PA
CBHW052317220526
45472CB00001B/152